高等职业教育大数据工程技术系列教材

大数据分析及应用项目教程
（Spark SQL）

许　慧　主　编
王张夫　杨　琳　副主编

电子工业出版社
Publishing House of Electronics Industry
北京·BEIJING

内容简介

本书以项目任务为载体串联知识与技能，介绍基于 Spark SQL 技术的大数据分析方法，培养学习者使用 Spark SQL 技术解决实际数据分析问题的能力，主要内容有大数据分析概述、实践环境准备、学生信息处理分析、房产大数据分析与探索、电商大数据分析与探索、Zepplin 数据可视化。在实现任务的过程中详细阐述了 Scala 语言基础、Spark SQL 编程分析、Spark SQL 数据分析与探索、数据可视化等知识。

本书适合作为高等职业院校计算机、大数据、人工智能及相关专业的教材或参考书，也可以作为数据分析人员的参考书。

未经许可，不得以任何方式复制或抄袭本书之部分或全部内容。
版权所有，侵权必究。

图书在版编目（CIP）数据

大数据分析及应用项目教程：Spark SQL / 许慧主编 . —北京：电子工业出版社，2023.9
高等职业教育大数据工程技术系列教材

ISBN 978-7-121-46363-1

Ⅰ.①大… Ⅱ.①许… Ⅲ.①数据处理—高等职业教育—教材 Ⅳ.① TP274

中国国家版本馆 CIP 数据核字（2023）第 175480 号

责任编辑：杨永毅
印　　刷：河北鑫兆源印刷有限公司
装　　订：河北鑫兆源印刷有限公司
出版发行：电子工业出版社
　　　　　北京市海淀区万寿路 173 信箱　　邮编 100036
开　　本：787×1 092　　1/16　　印张：14　　字数：341 千字
版　　次：2023 年 9 月第 1 版
印　　次：2023 年 9 月第 1 次印刷
印　　数：1200 册　　定价：55.00 元

凡所购买电子工业出版社图书有缺损问题，请向购买书店调换。若书店售缺，请与本社发行部联系，联系及邮购电话：（010）88254888，88258888。
质量投诉请发邮件至 zlts@phei.com.cn，盗版侵权举报请发邮件至 dbqq@phei.com.cn。
本书咨询联系方式：（010）88254570，xujj@phei.com.cn。

PREFACE 前言

当前"大数据"已经渗透到人们生活、生产、学习等领域的各个角落中,数据量呈现爆发式增长。"数据不是黄金,数据指引黄金",数据中蕴含的价值被广泛关注,大数据分析越来越流行。Spark SQL 作为 Spark 的核心技术之一,提供了一个名为 DataFrame 的编程抽象,可以充当分布式 SQL 查询引擎,为大规模数据处理分析提供技术支持。

本书以项目任务为载体串联知识与技能,并对其进行讲解与实践操作,共分为 6 章。第 1 章通过"关于大数据分析"和"认识 Spark SQL"两个任务介绍大数据分析的相关概念及 Spark SQL 的基础知识,使学习者对本书的内容有大致的了解。第 2 章通过"Hadoop 集群环境搭建"、"Spark 集群部署与使用"和"安装 Scala"任务,为全书实践环境及后续的数据分析做准备。第 3 章通过"班级基本情况分析"和"学生基本情况分析"两个任务对 Scala 语言在数据分析中的应用展开讲解,为之后的数据分析编程打基础。第 4 章通过"某房产公司销售人员业绩分析"和"某城市近年房产销售状况分析"两个任务讲解 Spark SQL 各种操作和各个算子在数据分析中的使用方法。第 5 章通过"'女装电子商务评论'数据分析"和"'在线销售订单'数据分析"两个任务讲解数据准备、清洗、转换、分析、输出、查询过程中用到的 DataFrame 操作方法。第 6 章通过对第 5 章两个任务的数据可视化,介绍基于 Web 的 Notebook 数据可视化工具——Zepplin。对于没有大数据技术基础或者基础薄弱的学习者,可以通过本书快速掌握大数据相关集群环境的搭建部署和利用 Spark SQL 进行数据分析。

本书设计了课程资源(视频资源、PPT、源码、习题及答案、软件安装包,搭建好的 Hadoop 集群、Spark 集群虚拟机镜像文件等)、课堂任务驱动设计、课后实验实训的实施体系,为学习者提供了完整的学习路线。

本书可以作为大数据等相关专业的教材,授课内容和学时安排如下表所示。其中"第 2 章 实践环境准备"内容可以拓展为"大数据集群环境搭建实训"的内容。

授课内容与学时安排

序 号	内 容	建 议 学 时
1	第 1 章 大数据分析概述	4
2	第 2 章 实践环境准备	6
3	第 3 章 学生信息处理分析	16
4	第 4 章 房产大数据分析与探索	16

续表

序 号	内 容	建 议 学 时
5	第 5 章 电商大数据分析与探索	14
6	第 6 章 Zepplin 数据可视化	8
	合计	64

　　本书注重思政育人，深入学习贯彻党的二十大精神，挖掘并融入素质元素，每个项目任务的开篇位置都有"素质目标"栏目，体现思政与素质教学目标。

　　此外，本书所有章节均先进行"情景导入"，提出"学习目标和要求"，再进行任务分析、讲解，编程实践操作，脑图小结巩固，章节练习提升。同时，本书提供大量翔实的源代码和操作步骤的图文，提高学习者对大数据分析相关技术的学习效率。

　　本书由衢州职业技术学院信息工程学院大数据技术专业教师编写，由许慧担任主编，由王张夫、杨琳担任副主编。其中，第 1 章由杨琳编写，第 2、3、4、5 章由许慧编写，第 6 章由王张夫编写，全书由许慧统稿。本书的编写得到了 2022 年浙江省"课程思政"示范课程项目（项目序号：646）的资助。

　　为方便学习者使用数字资源，本书中嵌套了对应数字资源的二维码，可扫描书中相应章节的二维码浏览学习。

　　为了方便教师教学，本书配有教学课件、课程大纲、授课计划、教案等相关资源，请有此需要的教师登录华信教育资源网注册后免费下载，如有问题可在网站留言板留言或与电子工业出版社联系（E-mail：hxedu@phei.com.cn）。

　　教材建设是一项系统工程，需要在实践中不断加以完善及改进，由于时间仓促、编者水平有限，书中难免存在疏漏和不足之处，敬请广大读者给予批评和指正。

编　者

CONTENTS 目录

第 1 章 大数据分析概述 1

任务 1.1 关于大数据分析 1
  情境导入 1
  学习目标和要求 1
  1.1.1 什么是大数据分析 2
  1.1.2 大数据分析工具 5
  1.1.3 大数据分析可视化 8

任务 1.2 认识 Spark SQL 8
  情境导入 8
  学习目标和要求 8
  1.2.1 Spark SQL 的背景简介 9
  1.2.2 Spark SQL 的运行原理 10

脑图小结 14
章节练习 16

第 2 章 实践环境准备 17

任务 2.1 Hadoop 集群环境搭建 17
  情境导入 17
  学习目标和要求 18
  2.1.1 环境准备 18
  2.1.2 安装 Hadoop 35
  2.1.3 启动 Hadoop 集群 45
  2.1.4 运行经典案例 wordcount 48

任务 2.2 Spark 集群部署与使用 50
  情境导入 50
  学习目标和要求 50
  2.2.1 安装 Spark 50
  2.2.2 启动 Spark 54

2.2.3　Spark 集群测试 56
　任务 2.3　安装 Scala 60
　　　情境导入 60
　　　学习目标和要求 60
　　　2.3.1　下载 Scala 安装包 61
　　　2.3.2　Scala 安装配置 61
　脑图小结 62
　章节练习 63

第 3 章　学生信息处理分析 65

　任务 3.1　班级基本情况分析 66
　　　情境导入 66
　　　学习目标和要求 66
　　　3.1.1　学生所属班级和男女生数量 67
　　　3.1.2　以班级为单位整理学生信息 84
　任务 3.2　学生基本情况分析 96
　　　情境导入 96
　　　学习目标和要求 96
　　　3.2.1　学生特长情况分析 96
　　　3.2.2　学生成绩情况分析 102
　脑图小结 105
　章节练习 106

第 4 章　房产大数据分析与探索 108

　任务 4.1　某房产公司销售人员业绩分析 108
　　　情境导入 108
　　　学习目标和要求 110
　　　4.1.1　数据集处理 110
　　　4.1.2　数据操作分析 116
　任务 4.2　某城市近年房产销售状况分析 131
　　　情境导入 131
　　　学习目标和要求 132
　　　4.2.1　数据准备 132
　　　4.2.2　数据探索与分析 134
　　　4.2.3　总结分析 139
　脑图小结 139
　章节练习 140

目录

第 5 章 电商大数据分析与探索 141

任务 5.1 "女装电子商务评论"数据分析 141
情境导入 141
学习目标和要求 142
5.1.1 数据准备 142
5.1.2 数据清洗 152
5.1.3 数据转换 157
5.1.4 数据分析 162
5.1.5 数据输出 171

任务 5.2 "在线销售订单"数据分析 173
情境导入 173
学习目标和要求 174
5.2.1 数据查询操作 174
5.2.2 数据分析探索 184

脑图小结 191
章节练习 192

第 6 章 Zeppelin 数据可视化 193

任务 6.1 Zeppelin 安装与部署 193
情境导入 193
学习目标和要求 193
6.1.1 下载安装包 194
6.1.2 安装配置 195
6.1.3 测试运行 Zeppelin 199

任务 6.2 "女装电子商务评论"数据可视化 203
情境导入 203
学习目标和要求 204
6.2.1 加载数据注册视图 204
6.2.2 执行 SQL 数据可视化 208

任务 6.3 "在线销售订单"数据可视化 212
情境导入 212
学习目标和要求 213
执行 Spark SQL 数据可视化 213

脑图小结 216
章节练习 216

第1章 大数据分析概述

任务 1.1 关于大数据分析

关于大数据分析

情境导入

李雷是一名大数据技术专业的大二学生,对大数据的相关技术内容非常感兴趣,在学习了大数据基础与运维、Python 编程基础、MySQL 数据库等知识之后,想对大数据分析进行更深一步的了解。他在初次接触大数据分析时,有许多不解之处,希望能得到大家的帮助。请完成本节内容的学习,掌握相关知识,为更好地掌握大数据分析基础理论知识做好准备。

学习目标和要求

知识与技能目标

1. 了解大数据分析的概念、特点、类别、优缺点。
2. 知道大数据分析的相关工具。
3. 了解大数据分析可视化的概念及相关工具。

素质目标

1. 具有数据分析意识与思维。
2. 提升大数据安全意识,树立遵守法律法规的观念。
3. 增强科技报国、不断奋斗的理想和信念。

1.1.1 什么是大数据分析

随着互联网技术、数据采集技术、硬件条件的发展及数据应用场景的扩展，物联网、互联网、移动智能终端等平台设施中产生了大量的数据。

众所周知，比较经典且被广泛认可的大数据的概念是 IBM 提出的"5V"特征，即 Volume（容量），数据存量庞大；Velocity（速度），数据的增长速度飞快；Variety（种类），数据类型千差万别；Value（价值），数据中蕴含可挖掘分析的价值；Veracity（真实性），数据的质量存在不确定性。在以上 5 个特征中，Value 是大数据的最终意义，即从数据中洞察价值。由此，大数据分析应运而生。

1. 大数据分析的概念

大数据是指无法在一定时间范围内用常规软件工具进行捕捉、管理和处理的数据集合，大数据分析就是指对规模巨大的数据进行分析。这些不同的数据集合包含来源不同的结构化、半结构化和非结构化数据，其大小从 TB 到 ZB 不等。数据分析是为了提取有用信息并形成结论，对数据加以详细研究和概括总结的过程。而大数据分析是使用高级分析技术分析大量不同的数据集，从这些原始数据中提取有意义的见解的过程。

2. 大数据分析的特点

1) 数据分析量大

大数据的数据量大决定了待处理分析的数据规模大。大数据中的数据源于互联网、物联网和传统信息系统等渠道。但对庞大的数据量的界定存在多种观点。比如有的观点认为当数据超过几十万、几百万条时就可以认为是很大的数据量；还有观点认为当数据量达到传统方法根本无法处理的程度时才算大数据，其中，非结构化数据处理维度很大，当达到上万条的数据量时就很难处理了。这些都可以称为大数据，即因为数据量庞大，所以在进行数据分析时需要借助其他工具和框架处理的数据。

2) 数据处理速度快

大数据的增长速度快，对大数据分析的实时性提出了要求。传统信息系统的数据增量通常是可以预测的，或者增长率是可控的；但是大数据时代的数据，其增长率已经远远超过了传统数据，对数据的处理能力已经超出了传统方法的极限。数据增长是一个相对的概念，与消费互联网相比，工业互联网带来的数据增长可能更加客观，因此工业互联网时代将进一步打开大数据的价值空间。

3) 数据分析类型多

结构化、半结构化、非结构化等繁多的大数据类型为数据处理增加了难度。相对于以往便于存储的、以文本为主的结构化数据，非结构化数据越来越多，包括网络日志、音频、视频、图片、地理位置信息等。这些多类型的数据对传统数据分析提出了巨大挑战，也是大数据分析兴起的重要原因。在工业互联网时代，数据结构的多样性将得到进一步体现，这也给数据价值提取过程带来了新的挑战。

4）数据价值密度低

数据量大、数据价值密度低是大数据的重要特征。而大数据的低价值密度使数据分析挖掘算法成了亟待解决的难题。传统数据基本都是结构化数据，每个字段都是有用的，价值密度非常高。与传统数据相比，大数据时代中越来越多的数据是半结构化和非结构化的，比如网站访问日志，其中大量内容都是没有价值的，真正有价值的内容比较少，虽然数据量比以前大了 N 倍，但价值密度却低了很多。这就需要更快、更方便的方式来完成数据价值的提取，也是当前大数据平台的核心竞争力之一。实际上，早期 Hadoop 平台和 Spark 平台能够脱颖而出的重要原因就是它们的数据处理、排序速度相对较快。

5）数据的可靠性低

大数据庞大的数据量一定会伴随着数据失真和部分数据不确定的情况，所以在分析过程中还要考虑数据的真实性问题。数据的可靠性是指在数据的生命周期内，所有数据都是完整、一致、准确的。保证数据的完整性需要以准确、真实、完全的方式来收集、记录、报告和保存实际产生的数据及信息。大数据带来的一个副作用是很难区分数据的真假，这也是当前大数据分析必须解决的重点问题。

3. 大数据分析的类型

大数据分析可以利用历史数据了解企业的历史情况，并为企业未来发展中的决策提供依据。大数据分析的常用分析类型有描述性分析、诊断性分析、预测性分析、规范性分析。

1）描述性分析

描述性分析是指描述过去的数据，基于大量历史数据进行汇总、分析、描述，并用简单可读的方式进行呈现。该类分析有利于进行总结汇报的创建。例如，利用回归技术从数据集中发现简单的趋势，利用可视化技术更直观地描述数据等。描述性分析通常被应用在商业智能系统等领域中。

2）诊断性分析

诊断性分析是指针对过去已经发生的事情分析其产生的原因。例如，对于电子商务公司客户流失的原因分析，可以根据商店情况、客户状态等历史数据进行数据恢复与分析挖掘。发现客户流失的原因可能是客服工作不到位、店铺优惠少、竞争店铺的广告效果更好等。再如，客户虽然将商品添加到了购物车中，但是商店的销售额还是有所下降，利用诊断性分析可以获知此问题背后的原因可能是订单没有正确加载、商品运费太高或者付款方式较少等。

在诊断分析的帮助下，企业可以获知问题的原因，并寻找解决手段。

3）预测性分析

顾名思义，预测性分析关注的是未来事件。利用历史数据和当前数据，通过统计分析、数据建模、机器学习等工具，对未来事件进行预测。这些未来事件可以是市场趋势、消费者行为趋势、服务提供商决策等相关的事件。预测性分析是企业中最常用的分析类型之一。

预测性分析并不能准确地告诉我们未来会发生什么，只能通过数据模型预测未来可能发生事情的概率。例如，跟踪历史数据记录，使用线性、回归等统计算法拟合数据趋势，预测未来的数据结果，并使用数据挖掘技术提取数据模式。

4）规范性分析

规范性分析是指在发现问题之后，根据问题诊断性分析和预测性分析，做出相应的优化建议和行动。规范性分析建立在描述性分析、诊断性分析和预测性分析3种类型的基础之上，根据合理的预测确定合适的执行方案。例如，一个公司的利润意外地激增或者下降，描述性分析和诊断性分析可以用于确定原因，预测性分析可以判断这种趋势在未来是否会延续，规范性分析则可以确定接下来的步骤，在把握机会的同时减少风险。

利用规范性分析，企业可以获取信息，从而为决策提供依据。例如，在企业网站、社交媒体、电子商务等平台上驱动推荐引擎，策划个性化内容。再如，在药物的开发过程中，为临床试验寻找合适的患者。又如，识别销售线索并将其转化为客户，为特定的客户提供相关服务等。

4. 大数据分析的优势与缺点

实践证明，大数据分析对于当今企业生产、生活服务具有重要意义与作用。大数据分析具有很多优势，同时也存在一定的缺点。

1）大数据分析的优势

（1）为优质决策提供参考。

用专业的大数据分析对企业各类数据进行分析挖掘，能够针对现实中的问题和需求为业务决策者提供决策依据与建议，推动业务竞争和提高发展效率；利用大数据分析目标市场需求，为客户提供量身定制的产品，减少低效的广告投入；利用大数据分析密切关注客户购买、交易等行为情况，帮助企业提高客户的品牌忠诚度。

（2）提高产品开发创新力。

企业越来越依赖市场洞察力来形成各种类型的业务实施战略。大数据分析可以帮助企业获知客户对产品的反馈，并利用这些数据来评估产品的性能，进而改善和创新产品，调整业务策略，评估产品产量等。

（3）改善客户服务体验。

利用大数据分析对从社交媒体、商业平台及其他渠道获取的客户各方面的数据进行挖掘分析，预测客户的习惯与喜好，从而提供贴合客户特征的产品或服务，极大地提高和改善客户的服务体验。

（4）提升风险管理。

企业可以利用大数据分析来预见任何可能发生的风险，并通过规范性分析与其他类型的分析技术来降低风险。同时，大数据分析还能提供有关消费者行为和市场趋势的重要意见，帮助企业评估其在市场中的地位和发展方向。

2）大数据分析的缺点

（1）信息透明化。

在大数据的环境下，数据比我们更了解我们自己。大数据的共享性质使得用户的数据容易泄露，从而造成许多的困扰和不便。信息的透明化必将导致网络安全风险，因此在进行大数据分析及相关操作的过程中必须遵守政府法律法规。

(2) 成本高。

大数据分析所需的时间很长，成本较高，具体体现在以下两方面。

第一，虽然当下许多大数据分析工具依赖开源技术能降低软件成本，但是企业在人员配备、硬件、系统维护等方面依然面临大量的开支。比如，支持大数据分析所需的互联网技术基础设施、用于存储大量数据的设备空间、用于数据传输的网络带宽，以及用于执行这些分析的计算资源，在购买和维护方面都十分昂贵。

第二，对很多已经经营多年的企业来说，在各种业务环境、应用程序和系统中已经存储了大量的数据，进行大数据分析就需要将以上这些不同类型的数据集合成为一体，并移动到指定的位置，这将大大增加处理的时间和费用。

(3) 数据质量低。

大数据分析的一个缺点是数据价值密度低，数据质量不高。如果不能解决数据质量的问题，则会使数据分析工作失去意义，甚至产生副作用。由于数据集庞大且复杂，因此在数据分析过程中，数据分析人员需要对数据进行处理转换等操作，以最大限度地保障数据的可靠性。

(4) 技术更新快。

大数据分析的飞速发展导致数据组织与应用工具非常多变，这也会增加数据分析人员的学习成本。

1.1.2 大数据分析工具

基于大数据的数据量大、速度增长快、数据种类多等特点，在不借助大数据平台的情况下，我们往往很难直接使用传统的分析工具来处理这些数据。例如，使用 Excel 等数据处理和分析工具，当单台计算机的数据吞吐量增加时，就会发现计算机极易发生卡顿并需要较长的响应时间，这是由计算机本身的计算逻辑决定的。因此，在数据量不断增大的同时，许多大数据分析工具与软件应运而生，并且随着技术的发展不断改进。

大数据分析工具类型众多，下面主要介绍 Hadoop 生态圈中的大数据分析工具、大数据分析编程语言和其他几种常用工具，以及对应的特点与优势。

1. Hadoop 生态圈中的大数据分析工具

Hadoop 是一个由 Apache 基金会开发的分布式系统基础架构，是目前领先的大数据分析开源工具之一。它主要包括 Hadoop 分布式文件系统（Hadoop Distributed File System，HDFS）、MapReduce 分布式计算系统和 YARN 分布式资源管理系统。用户可以在不了解分布式系统底层细节的情况下开发分布式程序，并充分利用集群的分布式功能进行运算和存储。

Hadoop 在大数据分析中具有可靠、高效、可伸缩的优点。其可靠性体现在维护多个数据副本时，可以避免计算和存储元素失败，并能够针对失败的节点重新进行分布式处理；其高效性体现在以并行的方式工作上，可通过并行处理加快处理速度；其可伸缩性体现在支持横向扩展上,可以很容易地增加和减少节点。此外,Hadoop 依赖于社区服务器,因此它的成本比较低,任何人都可以使用。

Hadoop 生态圈中的大数据分析工具有很多,下面简单介绍离线分析工具 Spark、HBase,实时分析工具 Storm、Flink。

1)Spark

Spark 是最好、最强大的开源大数据分析工具之一,数据分析人员可以借助其数据处理框架处理大量数据集。通过结合其他分布式计算工具,Spark 可以轻松地实现在多台计算机上分发数据处理任务。

Spark 具有 Spark SQL、Streaming 实时计算、Spark 机器学习和 SparkGraphX 图计算的内置功能,是大数据转换的快速和通用的生成器之一。Spark SQL 是用于结构化数据处理的 Spark 模块。Streaming 实时计算是一个基于 Spark Core 的实时计算框架,可以对大量数据源进行实时处理,并将结果存储到 HDFS、数据库中,具有高吞吐量和高容错的特点。Spark 机器学习提供了 MLlib 算法库,旨在简化机器学习的实践工作。SparkGraphX 图计算是一个分布式图处理框架,可以极大地满足数据分析人员对分布式图处理的需求。

与 MapReduce 相比,Spark 将数据存放于内存中,能以快 100 倍的速度处理数据。除此之外,Spark 还拥有丰富的高级算子,可以更快地构建并行应用程序。它还提供了 Java 中的高级 API,适用于不同的数据存储工具,如 HDFS、OpenStack 和 Cassandra 等,具有极大的灵活性和极强的多功能性。

2)HBase

HBase 是一个基于 HDFS 的面向列的分布式数据库,具有可伸缩、高可靠、高性能、分布式和面向列的特点。HBase 不同于一般的关系数据库,它是一个适合非结构化数据存储的数据库,采用基于列的而不是基于行的模式。HBase 中保存的数据可以使用 MapReduce 来处理,从而将数据存储和并行计算完美地结合在一起。

3)Storm

Storm 是 Twitter 开源的分布式实时大数据处理框架,可以简单、可靠地处理大量的数据流,如传感器数据、网站用户活动数据、金融交易数据等,随着时间的推移数据会源源不断地产生。流处理的基本职责是确保数据有效流动,同时提供可扩展和容错功能,Storm 就是流处理的代表之一。Storm 具有低延迟、高性能、分布式、可扩展、高容错、可靠、快速等特点。Storm 的计算应用场景有实时推荐系统、实时告警系统、股票系统、实时分析、在线机器学习、持续计算等。使用 Storm 的知名企业包括 Groupon 团购网站、淘宝、支付宝、阿里巴巴等。

4)Flink

Flink 是一个框架和分布式处理引擎,用于在无边界和有边界数据流上进行有状态的计算。无边界数据流没有时间的界限,所处理的数据是持续不断输入的。有边界数据流有时间的界限,比如我们常说的某天的数据、某月的数据。Flink 能在所有常见集群环境中运行,并能以内存速度和任意规模进行计算,具有高吞吐量、低延迟、高性能等特点。

2. 大数据分析编程语言

编程语言也是大数据分析的重要工具，下面介绍大数据分析中常用的编程语言。

1）Scala 语言

Scala 语言是 JVM 运行环境、面向对象和函数式编程的完美结合，是许多数据分析人员常用的语言。Scala 语言具有面向对象、函数式编程、可扩展和支持静态类型等特点。本书后续所有案例编程均使用 Scala 语言。

2）Python 语言

Python 语言在数据分析领域中是一个强大的工具。尽管入门难度相较于 BI、Excel 工具较高，但是在统计分析和预测分析等方面，Python 语言是其他工具无可比拟的。

3）R 语言

R 语言是大数据分析工具之一，是一种领先的统计编程语言，可用于科学计算、统计分析、数据可视化等。R 语言还可以扩展和集成其他数据分析工具，以执行各种大数据分析操作。

R 语言可以与任何编程语言集成，以提供更快的数据传输和更准确的分析。利用 R 语言，数据分析人员可以轻松地创建统计引擎，提供更精确的数据洞察力。此外，R 语言还提供了大量的绘图工具，便于数据分析和展示。

3. 其他常用工具

市面上还有许多大数据分析工具，可以应用于不同的场景。

1）RapidMiner

RapidMiner 是世界领先的数据分析解决方案，其特点是拖曳操作、无须编程、运算速度快，以及具有丰富的数据挖掘和算法功能。它常被用于解决各种商业关键问题，如营销响应率、客户细分、客户忠诚度及终身价值、资产维护、资源规划、预测性维修、质量管理、社交媒体监测和情感分析等。

2）MongoDB

MongoDB 是一个基于分布式文件存储的数据库，旨在为 Web 应用提供可扩展的高性能数据存储解决方案。它基于 NoSQL 数据库，可用于存储比基于 RDBMS 数据库更多的数据。

MongoDB 最大的特点是具有强大的查询语言，其语法类似面向对象的查询语言，几乎可以实现大部分类似关系数据库单表查询的功能，还支持对数据建立索引。

MongoDB 使用集合和文档进行操作，文档的基本数据单元是由键-值对（Key-Value）组成的，可以包含各种单元，但是大小、内容和字段数量因其中的文档而异。它允许数据分析人员基于编程语言中定义的类和对象更改文档结构。它还包含一个内置的数据模型，可以使数据分析人员利用层次关系来存储数组及其他元素。

以上工具只是众多大数据分析工具中的一部分，在具体的应用中还需要根据不同的数据类型与特点、数据分析问题与需求进行选择。

1.1.3 大数据分析可视化

数据可视化是指利用计算机，以图形、图表的形式将原始的抽象信息和数据直观地表示出来。利用图表、图形等元素可以形象地展现数据的趋势、规律，便于观察和理解。因此，大数据分析可视化就是利用各种工具和软件对大数据进行获取、清洗、挖掘、分析，并将分析结果通过图形、图表等形式进行呈现的过程。

与抽象的数字表格相比，大数据分析可视化的颜色和图案更容易吸引人们的目光和注意。因此，大数据分析可视化可以帮助我们从视觉角度对海量数据进行更加科学的阐释，并进一步引发观看者的兴趣，同时通过不同的表现方法与形式,将重点信息突出显示，使观看者的注意力集中于某一点上，从而使其能够容易理解并获取更加有价值的信息。

大数据分析可视化工具有很多，大致可以分为软件类和语言类。语言类的大数据分析可视化工具包括 Python 语言和 R 语言等。软件类的大数据分析可视化工具包括 Zeppelin、PowerBI、Tableau、Spass 等。本书将在第 6 章中使用基于 Web 的 Notebook 开发工具——Zeppelin 进行数据可视化展示。

任务 1.2　认识 Spark SQL

认识 Spark SQL

情境导入

李雷同学在学习了大数据分析基础知识后，理解了大数据分析的相关概念及重要意义，同时对大数据分析工具 Spark 中的 Spark SQL 产生了兴趣。Spark SQL 与传统 SQL 有什么区别和联系吗？它的基本架构、运行原理及流程是什么样的？ Spark SQL 的优势和特点是什么？请完成本节内容的学习，掌握相关知识，帮助李雷同学解决这些问题。

学习目标和要求

知识与技能目标

1. 了解 Spark SQL 的背景、特点。
2. 知道 Spark SQL 的运行架构。
3. 掌握 Catalyst 查询编译器的工作流程。
4. 掌握 Spark SQL 的运行流程。

素质目标
1. 具有强烈的好奇心和主动学习的职业品格。
2. 面对实际问题具有开拓创新的职业精神。

1.2.1 Spark SQL 的背景简介

1. Spark SQL 的背景

SQL 是结构化查询语言，具有专门为数据库建立的操作指令集，数据分析人员在使用时只需要发出指定命令，而不需要考虑具体的操作过程。SQL 的功能非常强大并且简单易学，已经成为数据库操作的基础。

随着大数据时代的飞速发展，数据规模不断扩大，传统的数据分析人员需要快速转换到大数据平台中。为了降低学习成本，使得数据分析人员能够直接在大数据平台上使用 SQL 进行上层的分析操作，SQL-on-Hadoop 工具应运而生。

Hive 是最原始的 SQL-on-Hadoop 工具，是由 Facebook 开发的构建于 Hadoop 集群之上的数据仓库应用。它提供了类似 SQL 语句的 HQL 语句作为数据访问接口，这使得传统数据分析人员应用 Hadoop 的学习曲线变缓。Hive 的本质是将 HQL 转换成 MapReduce 程序，来达到快速开发的目的。但是 MapReduce 计算过程中的 shuffle 过程是基于磁盘的，会消耗大量的 I/O，大大降低运行效率，运行复杂的 SQL ETL 经常需要数个小时，甚至数十个小时。为了提高 SQL-on-Hadoop 的效率，大量的工具开始出现，如 Apache Drill、Cloudera Impala、Shark 等。

Shark 是由加州大学伯克利分校研究人员使用 Scala 语言开发的开源 SQL 查询引擎，构建在已有的 Spark 数据处理引擎之上。它的设计目标是作为 Hive 的补充，同时在其工作节点上只执行查询操作，而不使用 MapReduce。Shark 底层使用 Spark 的基于内存的计算模型，可以使 SQL-on-Hadoop 的性能比 Hive 提高 10～100 倍。

实际上 Shark 和 Hive 是紧密关联的，Shark 虽然修改了内存管理、物理计划、执行 3 个模块，但是 Shark 对于 Hive 依旧存在很多的依赖，比如采用 Hive 的语法解析器、查询优化器等。

Shark 对 Hive 的依赖制约了 Spark 各个组件的相互集成，阻碍了它进一步发展。因此，一套全新的基于 Spark 框架的 Spark SQL 组件被开发并启用。Spark SQL 抛弃了 Shark 原有的代码，汲取了 Shark 的部分优点，比如内存列存储、Hive 兼容性等。在脱离了 Hive 的依赖之后，Spark SQL 在数据兼容、组件扩展、性能优化方面都得到了极大的提升。

2. Spark SQL 的特点

1）支持多种数据源

Spark SQL 可以在现有的 Hive 上运行 SQL 或者 HQL 进行查询操作。它不但兼容 Hive，还可以从弹性分布式数据集（Resilient Distributed Datasets，RDD）、Parquet 文件、JSON 文件中获取数据。

2）多种性能优化技术

Spark SQL 采用内存列存储（In-memory Columnar Storage）、字节码生成（Byte-code Generation）、Cost Model 技术对查询操作进行动态评估、获取最佳物理计划等。

Spark SQL 的表数据在内存中以内存列而非 JVM 对象存储，图 1-1 所示为两种存储方式的示意图。

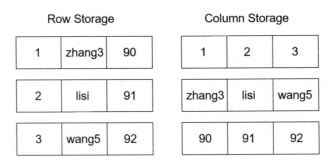

图 1-1　JVM 对象存储和内存列存储

在使用 JVM 对象存储方式时，每个对象通常都要增加 12～16 字节的额外开销，同时每个数据记录都会产生一个 JVM 对象。对于 GC（垃圾回收）来说，处理这么多的对象需要耗费更多的时间；对于基于内存计算的 Spark 来说，JVM 对象存储方式的代价太大。

内存列存储中数据集的每一列都作为一个数据存储单位，实现紧凑的数据存储，从而大大优化内存的使用效率。在采用此存储方式时，数据会更容易被读入内存并进行计算，对于分析查询中频繁使用的聚合特定列，性能会得到很大的提高。在采用了内存列存储之后，可以降低对内存的开销，从而避免回收大量数据产生的性能开销。

原生态的 JVM 内存模型会导致执行数据库查询语句中的表达式所需的时间成本变得非常昂贵。例如，执行"select a+b from table"命令，通用的 SQL 方法会先生成一个表达式，并多次调用虚函数，在这个过程中，会不断地打断 CPU 正常的流水线处理，减缓命令执行速度。因此，Spark SQL 在其 Catalyst 模块的 expressions 中增加了 Codegen 模块。对于 SQL 语句中的计算表达式，可以使用动态字节码生成技术来优化其性能，对匹配的表达式采用特定的代码动态编译并运行。

3）组件扩展性

用户可以对 SQL 的语法解析器、分析器及优化器进行重新定义和开发，并动态扩展。

1.2.2　Spark SQL 的运行原理

1. Spark SQL 的运行架构

在了解 Spark SQL 的运行原理之前，需要先认识 Spark SQL 的整体架构，如图 1-2 所示。Spark SQL 主要由 Catalyst、Core、Hive 和 Hive-Thriftserver 4 个子项目组成。

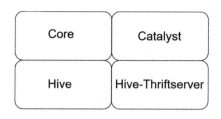

图 1-2　Spark SQL 的整体架构

1）Catalyst

Catalyst 是 Spark SQL 的可扩展查询优化系统，负责处理整个查询过程，包括解析、绑定、优化等，最后将 SQL 语句转换为物理计划（Physical Plan）。

2）Core

Core 是一个查询计划程序或者执行（planner/execution）的引擎，用于将 Catalyst 的逻辑查询计划转换为 Spark RDD 代码。Core 组件的公共接口 SQLContext 允许用户针对现有的 RDD 和 Parquet 文件执行 SQL 或 LINQ 语句。

3）Hive

Hive 组件包括 HiveContext 和 SQLContext，允许用户使用 HiveQL 的子集编写查询语句，并利用 Hive Metastore 从 Hive SerDe 中访问数据。Hive 还有一些允许用户运行包含 Hive UDF、UDAF 和 UDTF 查询的包装器。

4）Hive-Thriftserver

Hive-Thriftserver 支持 HiveServer 和 CLI，包括对 SQL CLI（bin/spark-sql）和 HiveServer2（用于 JDBC/ODBC）兼容服务器的支持。

2. Catalyst 查询编译器

在 Spark SQL 的各个组件中，Catalyst 查询编译器是最核心的部分。它负责语句的解析（Parse）、绑定、优化，根据逻辑计划生成物理计划等过程。Catalyst 的运行流程如图 1-3 所示。通过这一系列的操作将用户程序中的 SQL/DataFrame/Dataset 最终转化为在 Spark 平台上执行的 RDD。

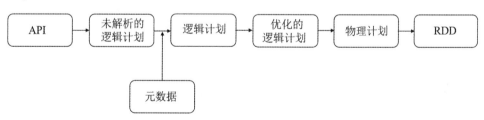

图 1-3　Catalyst 的运行流程

1）Catalyst 的组成

（1）Parser。

Parser 负责执行逻辑计划（Logical Plan），将 SQL/DataFrame/Dataset 转化为一棵未解析（Unresolved）的语法树。

（2）Analyzer。

Analyzer 中包含一系列的规则（Rule），每个规则负责不同的转换操作或者检查，比如解析 SQL 中的表名、列名，同时判断它们是否存在。Analyzer 会根据目录（Catalog）中的信息，对在 Parser 中生成的树进行解析，从而得到解析后的逻辑计划。

（3）Optimizer。

Optimizer 负责在逻辑优化阶段中对解析完成的逻辑计划应用一系列的规则以进行树结构的优化，从而获得更高的执行效率。这些规则包括常量合并、谓词下推、列裁切、布尔表达式简化等。

（4）Planner。

Planner 负责将优化的逻辑计划转换为物理计划。在物理计划的生成阶段，Spark SQL 会获取一个逻辑计划，并使用与 Spark 数据处理引擎匹配的物理操作来生成一个或多个物理计划。在转换过程中，一个逻辑计划可能对应多个物理计划的实现。例如，join 可以实现为 SortMergeJoin 或者 BroadcastHashJoin，此时需要使用基于代价的模型（Cost Model）在多个物理计划中选择最优的那一个。

整个 Catalyst 框架拥有良好的可扩展性，数据分析人员可以根据不同的需求，灵活地添加语法、解析规则、优化规则和转换策略。

2）Catalyst 的工作流程

在整个 Spark SQL 执行的过程中，Spark 会通过一些 API 接收 SQL 命令。在收到 SQL 命令之后，将其交给 Catalyst 进行解析，生成 RDD 的执行计划，最终交由集群执行。

在 Catalyst 中大致经过以下步骤。

首先，SQL 语句通过 Parser 模块被解析为语法树，该树被称为未解析的逻辑计划（Unresolved Logical Plan）。

然后，未解析的逻辑计划通过 Analyzer 模块并依据元数据被解析为逻辑计划。

接着，对各种基于规则的优化策略做进一步优化，得到优化的逻辑计划（Optimized Logical Plan）。

最后，优化的逻辑计划并不能被系统理解，因为其依然是逻辑的，所以需要将此逻辑计划转换为物理计划并交由集群执行。

3. Spark SQL 的运行流程

1）传统 SQL 的运行流程

明白传统关系型数据库的基本运行原理，有助于更好地理解 Spark SQL 的运行原理。传统 SQL 的运行流程如图 1-4 所示。

在传统关系型数据库中，最基本的 SQL 查询语句由投影（Projection）、数据源（DataSource）和过滤（Filter）3 部分组成，分别对应 SQL 查询过程中的 Result、DataSource 和 Operation，即 SQL 语法按照 Result → DataSource → Operation 的顺序来描述。但 SQL 的实际执行与 SQL 的语法是相反的，即按照 Operation → DataSource → Result 的顺序执行。

第 1 章 大数据分析概述

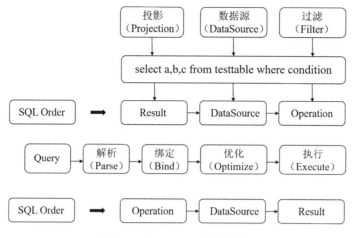

图 1-4 传统 SQL 的运行流程

详细的执行过程如下。

（1）词法和语法解析（Parse）：对写入的 SQL 命令进行词法和语法解析，分辨出命令中的关键词、表达式、投影、数据源等，判断 SQL 命令的语法是否规范，并形成逻辑计划。

（2）绑定（Bind）：将 SQL 命令与数据库的数据字典绑定，如果相关的投影和数据源等都存在，则表示这个 SQL 命令可以执行，从而生成可执行计划。

（3）优化（Optimize）：一条 SQL 查询命令可以有多种执行计划，优化器的最终目标是找到并生成最优的执行计划，交给执行引擎去执行。

传统关系型数据库的优化器是基于成本模型的，即根据已有的成本计算公式选择一个成本最低的执行方式。这个成本最低的执行方式不一定是最快的，但是在多数情况下是比较准确的。

（4）执行（Execute）：大部分核心事件已经被优化器处理完成，执行引擎只需要执行前面获取的最优执行计划，调用存储引擎的 API，查询数据返回得到的数据集即可。

2）Spark SQL 的运行流程

Spark SQL 的运行流程如图 1-5 所示。

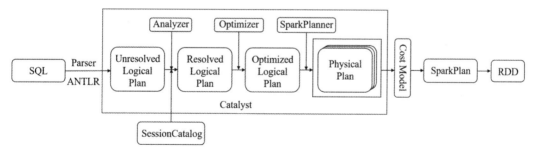

图 1-5 Spark SQL 的运行流程

Spark SQL 的运行分为以下几个步骤。

（1）SessionCatalog 保存元数据。

在解析 SQL 语句之前，需要初始化 SQLContext，创建 SparkSession，加载 SessionCatalog。在初始化 SQLContext 时，数据库名、表名、字段名称、字段类型等元数据都将被保存在 SessionCatalog 中。在逻辑计划分析器（Aralyzer）对未解析的逻辑计划进行处理时会使用这些数据。

（2）ANTLR 生成未解析的逻辑计划。

当调用 SparkSession 的 sql() 方法时，会使用 SparkSqlParser 对 SQL 语句进行解析，在解析过程中使用 ANTLR 进行词法解析和语法解析。Spark 从 2.0 版本开始使用 ANTLR 进行词法解析和语法解析，通过 ANTLR 构建一个按照关键字生成的语法树，即未解析的逻辑计划。

（3）Analyzer 解析逻辑计划。

在使用 Analyzer 分析器解析逻辑计划时，Analyzer 会使用 Analysis Rules，结合 SessionCatalog 元数据，对未解析的逻辑计划进行处理，并生成已解析的逻辑计划。

（4）Optimizer 优化逻辑计划。

Optimizer 会根据预先定义好的规则对已解析的逻辑计划进行迭代处理，优化并生成优化的逻辑计划。

（5）SparkPlanner 生成可执行的物理计划。

SparkPlanner 会使用 Planning Strategies，对优化的逻辑计划进行转换，生成多个可执行的物理计划。

（6）Cost Model 选择最佳物理计划。

根据过去的性能统计数据，Cost Model 会计算每个物理计划的代价，并选择代价最小的那一个生成最终的物理计划，得到 SparkPlan。

（7）QueryExecution 执行物理计划。

QueryExecution 会在执行物理计划之前，对 Preparations 进行规则处理，然后调用 SparkPlan 的 execute() 方法，返回 RDD。

通过上述步骤，就完成了从用户编写的 SQL 命令语句，到 Spark 内部 RDD 具体操作逻辑的转化。

脑图小结

本章首先对大数据分析进行了介绍，详细阐述了大数据分析的相关概念、特点、类别及优缺点；简单介绍了大数据分析的常用工具，并对大数据分析可视化的优势、价值等进行了分析。然后对大数据分析工具中的 Spark SQL 展开了详尽的描述，包括 Spark SQL 的发展演变历程、特点、运行架构、工作原理等内容。通过脑图小结，助力学习者掌握、巩固相关知识。

第1章 大数据分析概述

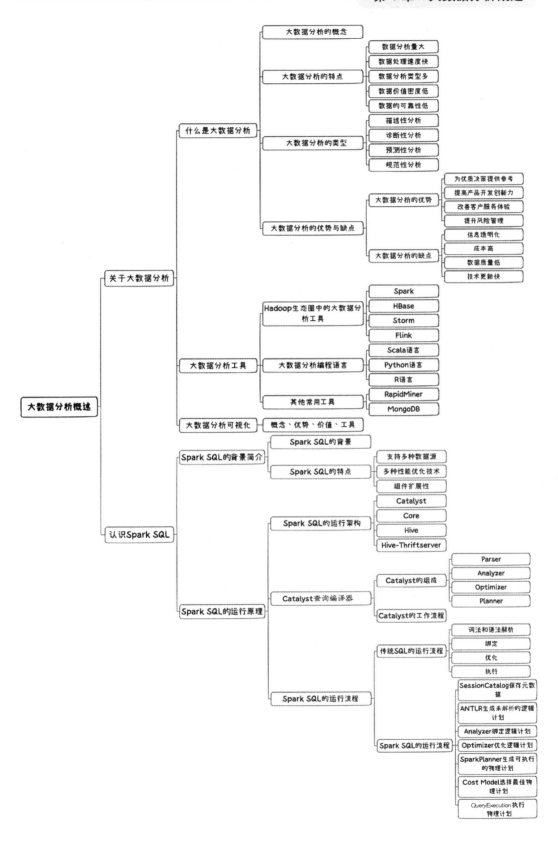

章节练习

1. 选择题

    (1) 以下不是大数据分析特征的是（　　）。
      A．数据价值密度高　　　　　　B．数据处理速度快
      C．数据分析类型多　　　　　　D．数据分析量大

    (2) 以下属于大数据分析类别的有（　　）。
      A．描述性分析　　　　　　　　B．诊断性分析
      C．预测性分析　　　　　　　　D．规范性分析

    (3) 以下不属于大数据分析工具的有（　　）。
      A．Hadoop　　　　　　　　　　B．Spark
      C．Illustrator　　　　　　　　　D．Spass

    (4) Spark SQL 的前身是（　　）。
      A．Shark　　　　　　　　　　 B．Cloudera Impala
      C．Apache Drill　　　　　　　 D．Hive

    (5) Spark SQL 的核心组件是（　　）。
      A．Catalyst　　　　　　　　　 B．Hive
      C．Core　　　　　　　　　　　D．Hive-Thriftserver

2. 填空题

    (1) 大数据的"5V"特征是指_____、_____、_____、_____、_____。

    (2) 大数据分析可视化就是利用各种工具和软件对大数据进行_____、_____、_____、_____，并将分析结果通过_____、_____等形式进行呈现的过程。

    (3) Spark SQL 的存储方式是_____。

    (4) Spark SQL 是由_____、_____、_____和_____4个子项目组成的。

    (5) Catalyst 是由_____、_____、_____、_____4个组件组成的。

3. 简答题

    (1) 大数据分析的优势有哪些？缺点又有哪些？

    (2) 简述 Spark SQL 的特点。

    (3) 简述 Spark SQL 的运行流程。

第 2 章

实践环境准备

任务 2.1　Hadoop 集群环境搭建

情境导入

Hadoop 是一个开源、可扩展、分布式的软件框架，它将计算任务以分布式的方式分配到集群的各个节点中进行处理，非常适合大数据业务的数据存储与计算。Hadoop 有 3 种部署模式，分别是单机模式、伪分布式模式、完全分布式模式。

1. 单机模式

单机模式是 Hadoop 的默认模式。当首次解压缩 Hadoop 的源码包时，Hadoop 无法了解硬件安装环境，便保守地选择了最小配置。这种默认模式下的 3 个 XML 配置文件均为空。当配置文件为空时，Hadoop 会完全在本地运行。单机模式下的主机不需要与其他节点交互，因此不会使用 HDFS，也不会加载任何 Hadoop 的守护进程。单机模式主要用于开发调试 MapReduce 程序的应用逻辑。

2. 伪分布式模式

伪分布式模式是指在一台主机上模拟多台主机，即 Hadoop 的守护进程在本地计算机上运行并模拟集群环境，它们是相互独立的 Java 进程。在这种模式下，Hadoop 使用的是分布式文件系统，各个作业也是由 JobTraker 服务管理的独立进程。伪分布式模式在单机模式的基础上增加了代码调试功能，允许检查内存使用情况，HDFS 的输入、输出，以及其他守护进程的交互，类似于完全分布式模式。因此，伪分布式模式常用来开发调试 Hadoop 程序的执行是否正确。

3. 完全分布式模式

完全分布式模式的守护进程运行在由多台主机搭建的集群上，是真正的生产环境。在所有的主机上安装 JDK 和 Hadoop，可以组成相互连通的网络。

在通常情况下，数据分析人员更希望以集群的方式运行 Spark，从而实现分布式计算。因此，在进行大数据分析之前，需要搭建一个 Hadoop 集群环境。搭建 Hadoop 集群环境需要先按步骤做好环境准备，再进行安装，接着按需求修改配置文件，最后初始化启动集群，从而完成整个集群环境的部署。整个过程的操作步骤在接下来的内容中将详细阐述，数据分析人员只需逐步执行即可实现 Hadoop 集群的搭建。

学习目标和要求

知识与技能目标

1. 掌握 Hadoop 集群环境搭建的准备工作，包括主机名配置、防火墙设置、免密登录设置、Java 环境配置。
2. 掌握安装 Hadoop、配置文件及启动集群的方法。
3. 能在 Hadoop 集群中运行经典案例 wordcount。

素质目标

1. 具有计算机系统的整体意识。
2. 养成大数据编程规范操作的工作习惯。

2.1.1 环境准备

Hadoop 集群环境搭建——环境准备

1. 集群节点规划

Hadoop 集群由 3 个节点构成，分别是 master、slaver01、slaver02，其 IP 地址和角色如表 2-1 所示。

表 2-1　集群节点规划

主　机　名	IP 地址	角　　色
master	192.168.128.130	ResourceManager SecondaryNameNode NameNode DataNode NodeManager
slaver01	192.168.128.131	DataNode NodeManager
slaver02	192.168.128.132	DataNode NodeManager

为了方便学习与操作，本书中集群的搭建和部署均在虚拟机中完成，使用 VMware Workstation 16 Pro 虚拟计算机软件。首先，准备 3 台拥有 CentOS 7 的虚拟机，可以通过本书提供的 BaseHost 虚拟机创建链接或者完全克隆获得，如图 2-1 所示，虚拟机登录密码设置为 6 个 0。

第 2 章 实践环境准备

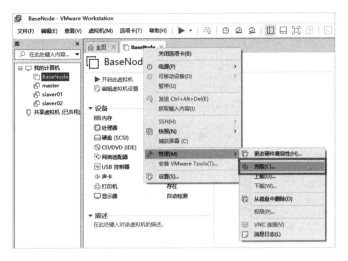

图 2-1 虚拟机克隆

2. 配置主机名和 IP 地址

为了区分 3 台虚拟机，需要对其主机名进行修改，分别命名为 master、slaver01、slaver02。同时需要修改 3 台虚拟机的 IP 地址，避免冲突。

1）配置 master 主机名、IP 地址

（1）修改网卡信息。

使用 vi 编辑器编辑网卡的信息，此处网卡名称为 ens33。若虚拟机是从本书提供的 BaseHost 节点中克隆而来的，则只需要修改 IP 地址即可。若虚拟机是新建的，则需要在修改 IP 地址之后添加子网掩码 NETMASK、DNS 和网关（GATEWAY）信息，还需要将网络配置参数 BOOTPROTO 修改为 static，即静态 IP 地址。

在第一台虚拟机中输入以下命令，修改网卡信息，结果如图 2-2 所示。

```
[root@localhost ~]# vi /etc/sysconfig/network-scripts/ifcfg-ens33
IPADDR=192.168.128.130
NETMASK=255.255.255.0
DNS1=114.114.114.114
GATEWAY=192.168.128.2
```

图 2-2 修改网卡信息（1）

19

（2）重启网络。

输入以下命令，重启网络并查看网卡信息，确认 IP 地址设置正确，结果如图 2-3 所示。

[root@localhost ~]# systemctl restart network
[root@localhost ~]# ip a

```
[root@localhost ~]# systemctl restart network
[root@localhost ~]# ip a
1: lo: <LOOPBACK,UP,LOWER_UP> mtu 65536 qdisc noqueue state UNKNOWN group default qlen 1000
    link/loopback 00:00:00:00:00:00 brd 00:00:00:00:00:00
    inet 127.0.0.1/8 scope host lo
       valid_lft forever preferred_lft forever
    inet6 ::1/128 scope host
       valid_lft forever preferred_lft forever
2: ens33: <BROADCAST,MULTICAST,UP,LOWER_UP> mtu 1500 qdisc pfifo_fast state UP group default qlen 10
00
    link/ether 00:0c:29:4c:bd:c4 brd ff:ff:ff:ff:ff:ff
    inet 192.168.128.130/24 brd 192.168.128.255 scope global noprefixroute ens33
       valid_lft forever preferred_lft forever
    inet6 fe80::c1c9:d03f:e7b9:dae7/64 scope link noprefixroute
       valid_lft forever preferred_lft forever
[root@localhost ~]#
```

图 2-3　重启网络并查看网卡信息（1）

（3）修改主机名。

修改第一台虚拟机的主机名为 master。在第一台虚拟机中输入以下命令，进行主机名的修改，结果如图 2-4 所示。

[root@localhost ~]# hostnamectl set-hostname master
使主机名修改生效
[root@localhost ~]# bash
查看修改结果
[root@master ~]# hostname
master

```
[root@localhost ~]# hostnamectl set-hostname master
[root@localhost ~]# bash
[root@master ~]#
```

图 2-4　修改主机名（1）

修改主机名的方式有临时修改和永久修改。临时修改主机名的命令是"hostname 临时主机名"。永久修改主机名的命令是 "hostnamectl set-hostname 永久主机名"，是对 "/etc/hostname" 文件内容进行的修改。

2）配置 slaver01 主机名、IP 地址

（1）修改网卡信息。

从节点的主机名和 IP 地址修改与主节点 master 一样。在第二台虚拟机中输入以下命令，结果如图 2-5 所示。

修改网卡信息
[root@localhost ~]# vi /etc/sysconfig/network-scripts/ifcfg-ens33
IPADDR=192.168.128.131
NETMASK=255.255.255.0
DNS1=114.114.114.114
GATEWAY=192.168.128.2

```
BROWSER_ONLY=no
BOOTPROTO=static
DEFROUTE=yes
IPV4_FAILURE_FATAL=no
IPV6INIT=yes
IPV6_AUTOCONF=yes
IPV6_DEFROUTE=yes
IPV6_FAILURE_FATAL=no
IPV6_ADDR_GEN_MODE=stable-privacy
NAME=ens33
UUID=f5771722-0909-4aa4-a978-ff35610a215a
DEVICE=ens33
ONBOOT=yes
IPADDR=192.168.128.131
NETMASK=255.255.255.0
DNS1=114.114.114.114
GATEWAY=192.168.128.2
```

图 2-5　修改网卡信息（2）

（2）重启网络。

输入以下命令，重启网络并查看网卡信息，确认 IP 地址设置正确，结果如图 2-6 所示。

[root@localhost ~]# systemctl restart network
[root@localhost ~]# ip a

```
[root@localhost ~]# systemctl restart network
[root@localhost ~]# ip a
1: lo: <LOOPBACK,UP,LOWER_UP> mtu 65536 qdisc noqueue state UNKNOWN group default qlen 1000
    link/loopback 00:00:00:00:00:00 brd 00:00:00:00:00:00
    inet 127.0.0.1/8 scope host lo
       valid_lft forever preferred_lft forever
    inet6 ::1/128 scope host
       valid_lft forever preferred_lft forever
2: ens33: <BROADCAST,MULTICAST,UP,LOWER_UP> mtu 1500 qdisc pfifo_fast state UP group default qlen 10
00
    link/ether 00:0c:29:ed:d4:ea brd ff:ff:ff:ff:ff:ff
    inet 192.168.128.131/24 brd 192.168.128.255 scope global noprefixroute ens33
       valid_lft forever preferred_lft forever
    inet6 fe80::2e6b:4b52:500c:f224/64 scope link noprefixroute
       valid_lft forever preferred_lft forever
    inet6 fe80::c1c9:d03f:e7b9:dae7/64 scope link tentative noprefixroute dadfailed
       valid_lft forever preferred_lft forever
[root@localhost ~]#
```

图 2-6　重启网络并查看网卡信息（2）

（3）修改主机名。

修改第二台虚拟机的主机名，命令如下，结果如图 2-7 所示。

[root@localhost ~]# hostnamectl set-hostname slaver01
[root@localhost ~]# bash
[root@slaver01 ~]# hostname
slaver01

```
[root@localhost ~]# hostnamectl set-hostname slaver01
[root@localhost ~]# bash
[root@slaver01 ~]#
```

图 2-7　修改主机名（2）

3）配置 slaver02 主机名、IP 地址

（1）修改网卡信息。

在第三台虚拟机中输入以下命令，修改网卡信息，结果如图 2-8 所示。

```
[root@localhost ~]# vi /etc/sysconfig/network-scripts/ifcfg-ens33
IPADDR=192.168.128.132
NETMASK=255.255.255.0
DNS1=114.114.114.114
GATEWAY=192.168.128.2
```

图 2-8　修改网卡信息（3）

（2）重启网络。

输入以下命令，重启网络并查看网卡信息，确认 IP 地址设置正确，结果如图 2-9 所示。

```
[root@localhost ~]# systemctl restart network
[root@localhost ~]# ip a
```

图 2-9　重启网络并查看网卡信息（3）

（3）修改主机名。

修改第三台虚拟机的主机名，命令如下，结果如图 2-10 所示。

```
[root@localhost ~]# hostnamectl set-hostname slaver02
[root@localhost ~]# bash
[root@slaver02 ~]# hostname
slaver02
```

图 2-10　修改主机名（3）

3. 连接 MobaXterm 终端工具

在修改主机名和 IP 地址之后，为了后续操作方便，我们使用 MobaXterm 终端工具进行连接。MobaXterm 又名 MobaXVT，是一款增强型终端、X 服务器和 UNIX 命令集（GNU/Cygwin）工具箱，可以用更简单的方式处理远程作业。

（1）打开 MobaXterm 软件，选择"Session"选项，创建连接，如图 2-11 所示。

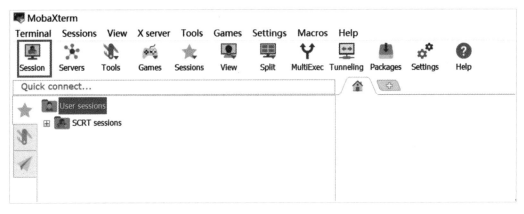

图 2-11　创建连接

（2）在"Session settings"对话框中选择"SSH"选项，输入虚拟机节点相关参数，在"Remote host"文本框中输入虚拟机的 IP 地址，勾选"Specify username"复选框并输入用户名"root"，Port（端口）默认为 22。除此之外，还可以设置终端的其他参数，比如修改外观等，最后点击"OK"按钮创建 SSH 连接，如图 2-12 所示，结果如图 2-13 所示。

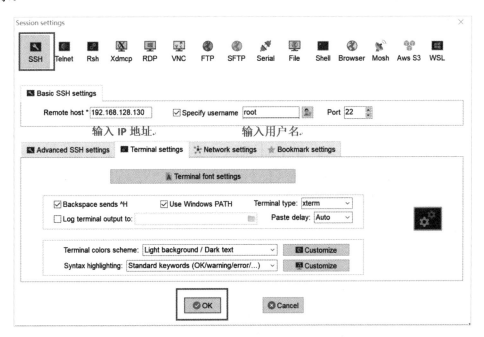

图 2-12　设置相关参数

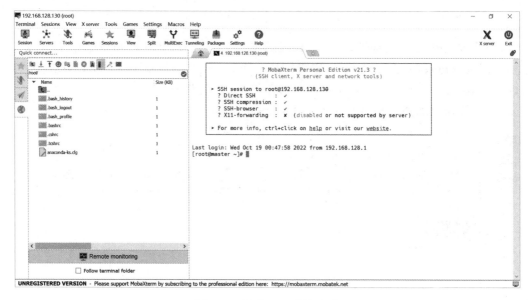

图 2-13 创建 SSH 连接完成

（3）为 slaver01 节点和 slaver02 节点创建 SSH 连接。

使用相同的方法连接另外两个节点，结果如图 2-14 所示。

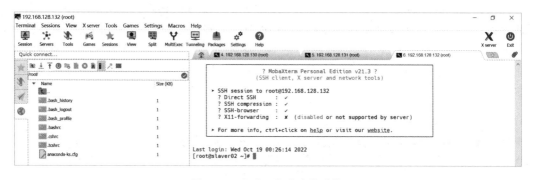

图 2-14 完成 3 个节点的连接

4. 关闭防火墙

防火墙是对服务器进行的一种保护服务，但有时候会带来很多麻烦，它会妨碍 Hadoop 集群间的相互通信，所以要将其关闭。注意，需要对 3 台虚拟机都进行防火墙关闭操作。

（1）关闭 master 节点防火墙并设置开机不自启。

在 master 节点中输入以下命令，结果如图 2-15 所示。

```
# 查看防火墙状态
[root@master ~]# systemctl status firewalld
# 关闭防火墙
[root@master ~]# systemctl stop firewalld
# 设置防火墙开机不自启
[root@master ~]# systemctl disable firewalld
```

再次查看防火墙状态
[root@master ~]# systemctl status firewalld

```
[root@master ~]# systemctl status firewalld
● firewalld.service - firewalld - dynamic firewall daemon
   Loaded: loaded (/usr/lib/systemd/system/firewalld.service; enabled; vendor preset: enabled)
   Active: active (running) since Tue 2022-10-18 23:24:59 EDT; 1h 41min ago
     Docs: man:firewalld(1)
 Main PID: 681 (firewalld)
   CGroup: /system.slice/firewalld.service
           └─681 /usr/bin/python -Es /usr/sbin/firewalld --nofork --nopid

Oct 18 23:24:58 localhost.localdomain systemd[1]: Starting firewalld - dynamic firewall daemon...
Oct 18 23:24:59 localhost.localdomain systemd[1]: Started firewalld - dynamic firewall daemon.
[root@master ~]# systemctl stop firewalld
[root@master ~]# systemctl disable firewalld
Removed symlink /etc/systemd/system/multi-user.target.wants/firewalld.service.
Removed symlink /etc/systemd/system/dbus-org.fedoraproject.FirewallD1.service.
[root@master ~]# systemctl status firewalld
● firewalld.service - firewalld - dynamic firewall daemon
   Loaded: loaded (/usr/lib/systemd/system/firewalld.service; disabled; vendor preset: enabled)
   Active: inactive (dead)
     Docs: man:firewalld(1)

Oct 18 23:24:58 localhost.localdomain systemd[1]: Starting firewalld - dynamic firewall daemon...
Oct 18 23:24:59 localhost.localdomain systemd[1]: Started firewalld - dynamic firewall daemon.
Oct 19 01:07:11 master systemd[1]: Stopping firewalld - dynamic firewall daemon...
Oct 19 01:07:11 master systemd[1]: Stopped firewalld - dynamic firewall daemon.
[root@master ~]#
```

图 2-15　关闭 master 节点防火墙并设置开机不自启

（2）关闭 slaver01 节点防火墙并设置开机不自启。

在 slaver01 节点中输入以下命令，结果如图 2-16 所示。

查看防火墙状态
[root@slaver01 ~]# systemctl status firewalld
关闭防火墙
[root@slaver01 ~]# systemctl stop firewalld
设置防火墙开机不自启
[root@slaver01 ~]# systemctl disable firewalld
再次查看防火墙状态
[root@slaver01 ~]# systemctl status firewalld

```
[root@slaver01 ~]# systemctl status firewalld
● firewalld.service - firewalld - dynamic firewall daemon
   Loaded: loaded (/usr/lib/systemd/system/firewalld.service; enabled; vendor preset: enabled)
   Active: active (running) since Wed 2022-10-19 00:18:00 EDT; 54min ago
     Docs: man:firewalld(1)
 Main PID: 672 (firewalld)
   CGroup: /system.slice/firewalld.service
           └─672 /usr/bin/python -Es /usr/sbin/firewalld --nofork --nopid

Oct 19 00:17:59 localhost.localdomain systemd[1]: Starting firewalld - dynamic firewall daemon...
Oct 19 00:18:00 localhost.localdomain systemd[1]: Started firewalld - dynamic firewall daemon.
[root@slaver01 ~]# systemctl stop firewalld
[root@slaver01 ~]# systemctl disable firewalld
Removed symlink /etc/systemd/system/multi-user.target.wants/firewalld.service.
Removed symlink /etc/systemd/system/dbus-org.fedoraproject.FirewallD1.service.
[root@slaver01 ~]# systemctl status firewalld
● firewalld.service - firewalld - dynamic firewall daemon
   Loaded: loaded (/usr/lib/systemd/system/firewalld.service; disabled; vendor preset: enabled)
   Active: inactive (dead)
     Docs: man:firewalld(1)

Oct 19 00:17:59 localhost.localdomain systemd[1]: Starting firewalld - dynamic firewall daemon...
Oct 19 00:18:00 localhost.localdomain systemd[1]: Started firewalld - dynamic firewall daemon.
Oct 19 01:12:36 slaver01 systemd[1]: Stopping firewalld - dynamic firewall daemon...
Oct 19 01:12:36 slaver01 systemd[1]: Stopped firewalld - dynamic firewall daemon.
[root@slaver01 ~]#
```

图 2-16　关闭 slaver01 节点防火墙并设置开机不自启

（3）关闭 slaver02 节点防火墙并设置开机不自启。

在 slaver02 节点中输入以下命令，结果如图 2-17 所示。

\# 查看防火墙状态
[root@slaver02 ~]# systemctl status firewalld
\# 关闭防火墙
[root@slaver02 ~]# systemctl stop firewalld
\# 设置防火墙开机不自启
[root@slaver02 ~]# systemctl disable firewalld
\# 再次查看防火墙状态
[root@slaver02 ~]# systemctl status firewalld

```
[root@slaver02 ~]# systemctl status firewalld
● firewalld.service - firewalld - dynamic firewall daemon
   Loaded: loaded (/usr/lib/systemd/system/firewalld.service; enabled; vendor preset: enabled)
   Active: active (running) since Wed 2022-10-19 00:25:57 EDT; 48min ago
     Docs: man:firewalld(1)
 Main PID: 684 (firewalld)
   CGroup: /system.slice/firewalld.service
           └─684 /usr/bin/python -Es /usr/sbin/firewalld --nofork --nopid

Oct 19 00:25:56 localhost.localdomain systemd[1]: Starting firewalld - dynamic firewall daemon...
Oct 19 00:25:57 localhost.localdomain systemd[1]: Started firewalld - dynamic firewall daemon.
[root@slaver02 ~]# systemctl stop firewalld
[root@slaver02 ~]# systemctl disable firewalld
Removed symlink /etc/systemd/system/multi-user.target.wants/firewalld.service.
Removed symlink /etc/systemd/system/dbus-org.fedoraproject.FirewallD1.service.
[root@slaver02 ~]# systemctl status firewalld
● firewalld.service - firewalld - dynamic firewall daemon
   Loaded: loaded (/usr/lib/systemd/system/firewalld.service; disabled; vendor preset: enabled)
   Active: inactive (dead)
     Docs: man:firewalld(1)

Oct 19 00:25:56 localhost.localdomain systemd[1]: Starting firewalld - dynamic firewall daemon...
Oct 19 00:25:57 localhost.localdomain systemd[1]: Started firewalld - dynamic firewall daemon.
Oct 19 01:14:27 slaver02 systemd[1]: Stopping firewalld - dynamic firewall daemon...
Oct 19 01:14:28 slaver02 systemd[1]: Stopped firewalld - dynamic firewall daemon.
[root@slaver02 ~]#
```

图 2-17　关闭 slaver02 节点防火墙并设置开机不自启

5. 关闭 SELinux

SELinux 是对系统安全级别更细粒度的设置，由于 SELinux 的配置太严格，可能会与其他服务产生冲突，所以这里需要将其关闭。注意，需要对 3 台虚拟机都进行 SELinux 关闭操作。

使用"getenforce"命令显示当前 SELinux 的应用模式，可以在"/etc/sysconfig/selinux"文件中修改 SELinux 的应用模式，选择 Enforcing（强制）、Permissive（宽容）或者 Disabled（关闭）。

Enforcing：强制模式，代表 SELinux 在运行中，且已经开始限制 domain/type 之间的验证关系。

Permissive：宽容模式，代表 SELinux 在运行中，但是不会限制 domain/type 之间的验证关系，即使验证不正确，进程仍可以对文件进行操作。如果验证不正确，则会发出警告。

Disabled：关闭模式，代表 SELinux 并没有实际运行。

（1）关闭 master 节点的 SELinux。

在 master 节点中输入以下命令，结果如图 2-18 所示。

\# 查看 SELinux 状态

```
[root@master ~]# getenforce
Enforcing
```
#setenforce 是 Linux 的 SELinux 防火墙配置命令，执行 setenforce 0 表示临时关闭 SELinux 防火墙
```
[root@master ~]# setenforce 0
```
临时关闭 SELinux
```
[root@master ~]# vi /etc/sysconfig/selinux
```

```
# This file controls the state of SELinux on the system.
# SELINUX= can take one of these three values:
#     enforcing - SELinux security policy is enforced.
#     permissive - SELinux prints warnings instead of enforcing.
#     disabled - No SELinux policy is loaded.
SELINUX=disabled  ←——从 enforcing 修改为 disabled
# SELINUXTYPE= can take one of three two values:
#     targeted - Targeted processes are protected,
#     minimum - Modification of targeted policy. Only selected processes are protected.
#     mls - Multi Level Security protection.
SELINUXTYPE=targeted
```

图 2-18　关闭 master 节点的 SELinux

（2）关闭 slaver01、slaver02 节点的 SELinux。

用同样的方法关闭另外两个节点的 SELinux，结果如图 2-19 和图 2-20 所示。

```
[root@slaver01 ~]# vi /etc/sysconfig/selinux
[root@slaver01 ~]# cat /etc/sysconfig/selinux

# This file controls the state of SELinux on the system.
# SELINUX= can take one of these three values:
#     enforcing - SELinux security policy is enforced.
#     permissive - SELinux prints warnings instead of enforcing.
#     disabled - No SELinux policy is loaded.
SELINUX=disabled
# SELINUXTYPE= can take one of three two values:
#     targeted - Targeted processes are protected,
#     minimum - Modification of targeted policy. Only selected processes are protected.
#     mls - Multi Level Security protection.
SELINUXTYPE=targeted
```

图 2-19　关闭 slaver01 节点的 SELinux

```
[root@slaver02 ~]# vi /etc/sysconfig/selinux
[root@slaver02 ~]# cat /etc/sysconfig/selinux

# This file controls the state of SELinux on the system.
# SELINUX= can take one of these three values:
#     enforcing - SELinux security policy is enforced.
#     permissive - SELinux prints warnings instead of enforcing.
#     disabled - No SELinux policy is loaded.
SELINUX=disabled
# SELINUXTYPE= can take one of three two values:
#     targeted - Targeted processes are protected,
#     minimum - Modification of targeted policy. Only selected processes are protected.
#     mls - Multi Level Security protection.
SELINUXTYPE=targeted
```

图 2-20　关闭 slaver02 节点的 SELinux

6．修改"hosts"文件

修改"hosts"文件，建立主机和 IP 地址之间的映射关系。"hosts"文件在 Linux 系统中可用于 IP 地址与域名快速解析。通过这个文件可以配置主机 IP 地址及对应的主机名，对于服务器类型的 Linux 系统的作用是不可忽略的。在局域网或 Internet 上，每台主机都有一个 IP 地址，可以用于区分各台主机，并根据 IP 地址进行通信。但 IP 地址不方便记忆，所以又有了域名。在一个局域网中，每台主机都有一个主机名，可以用于

区分主机，便于相互访问。

"hosts"文件可以解决远程登录 Linux 主机过慢的问题。有时客户端想远程登录一台 Linux 主机，但每次输入密码之后都会等很长一段时间才能进入，这是因为 Linux 主机在返回信息时需要解析 IP 地址，如果在 Linux 主机的"hosts"文件中事先加入客户端的 IP 地址，这时从客户端远程登录 Linux 就会变得很快。

在通常情况下，"hosts"文件会先记录本机的 IP 地址和主机名，其默认配置如图 2-21 所示。

```
127.0.0.1    localhost localhost.localdomain localhost4 localhost4.localdomain4
::1          localhost localhost.localdomain localhost6 localhost6.localdomain6
```

图 2-21 "hosts"文件默认配置

其默认配置格式如下。

主机 IP 地址 主机名

格式说明：在一般情况下，"hosts"文件的每行为一台主机的信息，即主机 IP 地址和主机名。

（1）修改 master 节点的"hosts"文件。

在 master 节点中输入以下命令，在"hosts"文件中添加 IP 地址和对应的主机名，结果如图 2-22 所示。

[root@master ~]# vi /etc/hosts

```
127.0.0.1    localhost localhost.localdomain localhost4 localhost4.localdomain4
::1          localhost localhost.localdomain localhost6 localhost6.localdomain6
192.168.128.130 master
192.168.128.131 slaver01
192.168.128.132 slaver02
~
```

图 2-22 "hosts"文件配置内容

（2）修改 slaver01 和 slaver02 节点的"hosts"文件。

由于另外两个节点的"hosts"文件和 master 节点是一样的，因此只需要将 master 节点中修改后的"hosts"文件发送到 slaver01、slaver02 节点中即可。

在 master 节点中输入以下命令，将"hosts"文件发送到 slaver01 节点中。

[root@master ~]# scp /etc/hosts root@slaver01:/etc/hosts

在命令执行过程中，由于还未设置免密登录，所以需要先输入"yes"确认传输文件，再输入 slaver01 的用户密码，此处密码是 000000，显示 100%（传输完成）即发送成功，结果如图 2-23 所示。

```
[root@master ~]# scp /etc/hosts root@slaver01:/etc/hosts
The authenticity of host 'slaver01 (192.168.128.131)' can't be established.
ECDSA key fingerprint is SHA256:1U8t89wlCTfGnrHZ7LangGx0LH/Lb9SL8aYAijC2Xu4.
ECDSA key fingerprint is MD5:c1:42:21:41:d6:94:56:2c:0b:5e:69:b3:86:d4:9d:d0.
Are you sure you want to continue connecting (yes/no)? yes   ← 输入 yes
Warning: Permanently added 'slaver01,192.168.128.131' (ECDSA) to the list of known hosts.
root@slaver01's password:              ← 输入密码 000000
hosts                                       100%  231   27.1KB/s   00:00
[root@master ~]#
```

图 2-23 将 master 节点的"hosts"文件发送到 slaver01 节点中

同理，在 master 节点中输入以下命令，将"hosts"文件发送到 slaver02 节点中，结果如图 2-24 所示。

[root@master ~]# scp /etc/hosts root@slaver02:/etc/hosts

```
[root@master ~]# scp /etc/hosts root@slaver02:/etc/hosts
The authenticity of host 'slaver02 (192.168.128.132)' can't be established.
ECDSA key fingerprint is SHA256:1U8t89wlCTfGnrHZ7LangGx0LH/Lb9SL8aYAijC2Xu4.
ECDSA key fingerprint is MD5:c1:42:21:41:d6:94:56:2c:0b:5e:69:b3:86:d4:9d:d0.
Are you sure you want to continue connecting (yes/no)? yes
Warning: Permanently added 'slaver02,192.168.128.132' (ECDSA) to the list of known hosts.
root@slaver02's password:
hosts                                             100%  231   145.0KB/s   00:00
[root@master ~]#
```

图 2-24　将 master 节点的"hosts"文件发送到 slaver02 节点中

（3）验证设置。

分别在 3 个节点中输入以下命令，查看 3 个节点之间是否能相互 ping 通，从而验证"hosts"文件修改是否正确。验证结果如图 2-25、图 2-26、图 2-27 所示。

ping master
ping slaver01
ping slaver02

```
[root@master ~]# ping slaver01
PING slaver01 (192.168.128.131) 56(84) bytes of data.
64 bytes from slaver01 (192.168.128.131): icmp_seq=1 ttl=64 time=1.27 ms
64 bytes from slaver01 (192.168.128.131): icmp_seq=2 ttl=64 time=2.72 ms
^Z
[2]+  Stopped                 ping slaver01
[root@master ~]# ping slaver02
PING slaver02 (192.168.128.132) 56(84) bytes of data.
64 bytes from slaver02 (192.168.128.132): icmp_seq=1 ttl=64 time=0.429 ms
64 bytes from slaver02 (192.168.128.132): icmp_seq=2 ttl=64 time=1.37 ms
^Z
[3]+  Stopped                 ping slaver02
[root@master ~]#
```

图 2-25　master 节点验证结果

```
[root@slaver01 ~]# ping master
PING master (192.168.128.130) 56(84) bytes of data.
64 bytes from master (192.168.128.130): icmp_seq=1 ttl=64 time=0.827 ms
64 bytes from master (192.168.128.130): icmp_seq=2 ttl=64 time=1.11 ms
^Z
[1]+  Stopped                 ping master
[root@slaver01 ~]# ping slaver01
PING slaver01 (192.168.128.131) 56(84) bytes of data.
64 bytes from slaver01 (192.168.128.131): icmp_seq=1 ttl=64 time=0.041 ms
64 bytes from slaver01 (192.168.128.131): icmp_seq=2 ttl=64 time=0.052 ms
^Z
[2]+  Stopped                 ping slaver01
[root@slaver01 ~]# ping slaver02
PING slaver02 (192.168.128.132) 56(84) bytes of data.
64 bytes from slaver02 (192.168.128.132): icmp_seq=1 ttl=64 time=0.761 ms
64 bytes from slaver02 (192.168.128.132): icmp_seq=2 ttl=64 time=0.853 ms
^Z
[3]+  Stopped                 ping slaver02
[root@slaver01 ~]#
```

图 2-26　slaver01 节点验证结果

```
[root@slaver02 ~]# ping master
PING master (192.168.128.130) 56(84) bytes of data.
64 bytes from master (192.168.128.130): icmp_seq=1 ttl=64 time=0.869 ms
64 bytes from master (192.168.128.130): icmp_seq=2 ttl=64 time=1.02 ms
^Z
[1]+  Stopped                 ping master
[root@slaver02 ~]# ping slaver01
PING slaver01 (192.168.128.131) 56(84) bytes of data.
64 bytes from slaver01 (192.168.128.131): icmp_seq=1 ttl=64 time=0.659 ms
64 bytes from slaver01 (192.168.128.131): icmp_seq=2 ttl=64 time=1.59 ms
^Z
[2]+  Stopped                 ping slaver01
[root@slaver02 ~]# ping slaver02
PING slaver02 (192.168.128.132) 56(84) bytes of data.
64 bytes from slaver02 (192.168.128.132): icmp_seq=1 ttl=64 time=0.045 ms
64 bytes from slaver02 (192.168.128.132): icmp_seq=2 ttl=64 time=0.069 ms
^Z
[3]+  Stopped                 ping slaver02
[root@slaver02 ~]#
```

图 2-27　slaver02 节点验证结果

由验证结果可知，3 个节点之间可以相互通信，表示"hosts"文件修改成功。

7. 配置免密登录

在准备好 Hadoop 集群之后，namenode 会通过 SSH 来启动和停止各个节点上的各种守护进程，这就需要以不输入密码的方式在节点之间执行命令，因此需要配置 SSH 使用无密码认证的方式。

首先在 master 节点中创建密钥，然后将密钥复制到 slaver01 和 slaver02 两个节点中，完成免密登录配置。

（1）在 master 节点中创建密钥。

在 master 节点中输入以下命令，一直按回车键直至结束，生成密钥，结果如图 2-28 所示。

[root@master ~]# ssh-keygen

```
[root@master ~]# ssh-keygen
Generating public/private rsa key pair.
Enter file in which to save the key (/root/.ssh/id_rsa):
Enter passphrase (empty for no passphrase):
Enter same passphrase again:
Your identification has been saved in /root/.ssh/id_rsa.
Your public key has been saved in /root/.ssh/id_rsa.pub.
The key fingerprint is:
SHA256:0ZWXunieO3xCZ7x8rjm/d+Iu7yBOm+ONFXStnCzqi08 root@master
The key's randomart image is:
+---[RSA 2048]----+
|              .. |
|          . .. o.|
|         . . .o. |
|          . ..+ o|
|         S  .oo= |
|           ..+o+ |
|           E=o= .|
|          *.BB.*.+|
|         .oXoo@BO*|
+----[SHA256]-----+
[root@master ~]#
```

图 2-28　生成密钥

（2）复制密钥到 master 节点中。

在复制密钥的过程中，需要先输入"yes"确认复制操作，然后输入密钥接收节点的用户密码，此处密码是 000000，操作过程如图 2-29 所示。

[root@master ~]# ssh-copy-id master
[root@master ~]# ssh-copy-id localhost

```
[root@master ~]# ssh-copy-id master
/usr/bin/ssh-copy-id: INFO: Source of key(s) to be installed: "/root/.ssh/id_rsa.pub"
The authenticity of host 'master (192.168.128.130)' can't be established.
ECDSA key fingerprint is SHA256:1U8t89wlCTfGnrHZ7LangGx0LH/Lb9SL8aYAijC2Xu4.
ECDSA key fingerprint is MD5:c1:42:21:41:d6:94:56:2c:0b:5e:69:b3:86:d4:9d:d0.
Are you sure you want to continue connecting (yes/no)? yes
/usr/bin/ssh-copy-id: INFO: attempting to log in with the new key(s), to filter out any that are alr
eady installed
/usr/bin/ssh-copy-id: INFO: 1 key(s) remain to be installed -- if you are prompted now it is to inst
all the new keys
root@master's password:                    ← 输入"yes"和密码

Number of key(s) added: 1

Now try logging into the machine, with:   "ssh 'master'"
and check to make sure that only the key(s) you wanted were added.

[root@master ~]# ssh-copy-id localhost
/usr/bin/ssh-copy-id: INFO: Source of key(s) to be installed: "/root/.ssh/id_rsa.pub"
The authenticity of host 'localhost (::1)' can't be established.
ECDSA key fingerprint is SHA256:1U8t89wlCTfGnrHZ7LangGx0LH/Lb9SL8aYAijC2Xu4.
ECDSA key fingerprint is MD5:c1:42:21:41:d6:94:56:2c:0b:5e:69:b3:86:d4:9d:d0.
Are you sure you want to continue connecting (yes/no)? yes
/usr/bin/ssh-copy-id: INFO: attempting to log in with the new key(s), to filter out any that are alr
eady installed
/usr/bin/ssh-copy-id: WARNING: All keys were skipped because they already exist on the remote system
.
              (if you think this is a mistake, you may want to use -f option)
```

图 2-29　复制密钥到 master 节点中

在 master 节点中验证免密登录，命令如下。

[root@master ~]# ssh master
[root@master ~]# ssh localhost

```
[root@master ~]# ssh master
Last login: Wed Oct 19 00:49:32 2022 from 192.168.128.1
[root@master ~]# ssh localhost
Last login: Wed Oct 19 02:20:50 2022 from master
[root@master ~]# logout
Connection to localhost closed.
[root@master ~]#
```

图 2-30　在 master 节点中验证免密登录

由图 2-30 可知，在设置免密登录之后，master 节点在使用 SSH 连接 master 和 localhost 时均无须密码。

（3）复制密钥到 slaver01 节点中。

在 master 节点中输入"ssh-copy-id slaver01"命令，将密钥复制到 slaver01 节点中，其间需要输入 slaver01 的密码"000000"确认发送。在完成密钥复制之后，输入"ssh slaver01"命令，从 master 节点登录到 slaver01 节点，验证免密登录是否配置成功。在验证免密登录成功之后，按"Ctrl+D"组合键退出登录，回到 master 节点中，操作过程如图 2-31 所示。

[root@master ~]# ssh-copy-id slaver01

[root@master ~]# ssh slaver01

```
[root@master ~]# ssh-copy-id slaver01
/usr/bin/ssh-copy-id: INFO: Source of key(s) to be installed: "/root/.ssh/id_rsa.pub"
/usr/bin/ssh-copy-id: INFO: attempting to log in with the new key(s), to filter out any that are alr
eady installed
/usr/bin/ssh-copy-id: INFO: 1 key(s) remain to be installed -- if you are prompted now it is to inst
all the new keys
root@slaver01's password:          输入密码
Number of key(s) added: 1

Now try logging into the machine, with:   "ssh 'slaver01'"
and check to make sure that only the key(s) you wanted were added.

[root@master ~]# ssh slaver01  ← 从 master 节点登录 slaver01，无须密码
Last login: Wed Oct 19 00:51:41 2022 from 192.168.128.1
[root@slaver01 ~]# logout  ← 按 "Ctrl+D" 组合键退出
Connection to slaver01 closed.
[root@master ~]#
```

图 2-31　复制密钥到 slaver01 节点中并验证免密登录

（4）复制密钥到 slaver02 节点中。

参照（3）中的操作，复制密钥到 slaver02 节点中，在 master 节点中输入以下命令，结果如图 2-32 所示。

[root@master ~]# ssh-copy-id slaver02
[root@master ~]# ssh slaver02

```
[root@master ~]# ssh-copy-id slaver02
/usr/bin/ssh-copy-id: INFO: Source of key(s) to be installed: "/root/.ssh/id_rsa.pub"
/usr/bin/ssh-copy-id: INFO: attempting to log in with the new key(s), to filter out any that are alr
eady installed
/usr/bin/ssh-copy-id: INFO: 1 key(s) remain to be installed -- if you are prompted now it is to inst
all the new keys
root@slaver02's password:
Number of key(s) added: 1

Now try logging into the machine, with:   "ssh 'slaver02'"
and check to make sure that only the key(s) you wanted were added.

[root@master ~]# ssh slaver02
Last login: Wed Oct 19 00:52:00 2022 from 192.168.128.1
[root@slaver02 ~]# logout
Connection to slaver02 closed.
[root@master ~]#
```

图 2-32　复制密钥到 slaver02 节点中并验证免密登录

8. 配置 Java 环境

Hadoop 采用 Java 编写，因此在安装之前需要配置 Java 环境。为了提高配置 Java 环境的效率，可以先在 master 节点中完成配置，再将相关的文件发送给另外两个节点。

1）配置 master 节点的 Java 环境

本书使用的 JDK 版本是 jdk1.8.0_281。使用 MobaXterm 工具，将软件包上传至 master 节点的 "root" 目录下，如图 2-33 所示。

第 2 章 实践环境准备

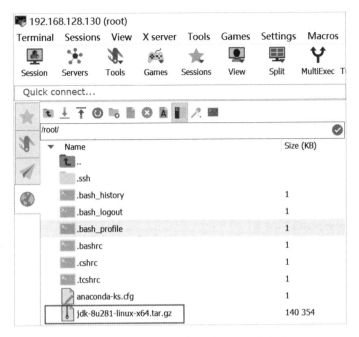

图 2-33　上传 JDK 软件包到 master 节点中

在完成软件包的上传之后，在 master 节点中输入以下命令，解压缩 JDK 软件包并查看结果，如图 2-34 所示。

[root@master ~]# tar -zxvf ./jdk-8u281-linux-x64.tar.gz -C /usr/local
[root@master ~]# ls /usr/local/

```
[root@master ~]# ls /usr/local/
bin  etc  games  include  jdk1.8.0_281  lib  lib64  libexec  sbin  share  src
[root@master ~]#
```

图 2-34　解压缩并查看 JDK 软件包

在解压缩安装包完成之后，需要在".bash_profile"文件中进行环境变量配置。在 master 节点中输入以下命令，修改".bash_profile"文件，添加内容如图 2-35 所示。

[root@master ~]# vi .bash_profile

图 2-35　配置 PATH 变量

在".bash_profile"文件中，使用 JDK 解压缩后的路径设置 JAVA_HOME 环境变量。添加 PATH 路径，PATH 环境变量的作用是在系统执行某条命令时，指引操作系统根据

33

它所指示的路径在系统中寻找该命令,告诉操作系统去哪里找这个命令文件,如果没有找到,则运行会报错。

在完成环境变量的配置之后,还需要使其生效,在 master 节点中输入"source .bash_profile"命令,使".bash_profile"文件的配置生效。查看 Java 版本以验证 Java 环境配置是否成功,若能正确显示 Java 版本,则表示环境配置成功,如图 2-36 所示。

[root@master ~]# source .bash_profile
[root@master ~]# java -version

图 2-36　环境配置成功

2)配置 slaver01、slaver02 节点的 Java 环境

slaver01、slaver02 节点的 Java 环境配置与 master 节点一样,因此只需要将 master 节点中的相关文件复制到另外两个节点中,并使".bash_profile"配置文件生效即可。

(1)在 master 节点中输入以下命令,将文件分发给 slaver01 节点。

分发 JDK 安装相关文件
[root@master ~]# scp -r /usr/local/jdk1.8.0_281/ root@slaver01:/usr/local
分发环境变量文件
[root@master ~]# scp .bash_profile root@slaver01:/root

在 slaver01 节点中输入以下命令,使环境变量生效并确认 Java 版本,结果如图 2-37 所示。

[root@slaver01 ~]# source .bash_profile
[root@slaver01 ~]# java -version

图 2-37　查看 slaver01 节点的 Java 版本

(2)在 master 节点中输入以下命令,分发文件到 slaver02 节点中。

分发 JDK 安装相关文件
[root@master ~]# scp -r /usr/local/jdk1.8.0_281/ root@slaver02:/usr/local
分发环境变量文件
[root@master ~]# scp .bash_profile root@slaver02:/root

在 slaver02 节点中输入以下命令,确认分发结果,安装成功,结果如图 2-38 所示。

[root@slaver02 ~]# source .bash_profile
[root@slaver02 ~]# java -version

```
[root@slaver02 ~]# source .bash_profile
[root@slaver02 ~]# java -version
java version "1.8.0_281"
Java(TM) SE Runtime Environment (build 1.8.0_281-b09)
Java HotSpot(TM) 64-Bit Server VM (build 25.281-b09, mixed mode)
[root@slaver02 ~]#
```

图 2-38　查看 slaver02 节点的 Java 版本

2.1.2　安装 Hadoop

安装 Hadoop

本书安装的 Hadoop 版本是 Hadoop-3.2.1。3 个节点都需要安装 Hadoop，为了提高部署效率，先在 master 节点中部署安装，再将相关的文件和配置复制分发到另外两个节点中。

1. 解压缩 Hadoop 软件包

（1）使用 MobaXterm 工具，将 Hadoop 软件包上传至 master 节点的"root"目录下，如图 2-39 所示。

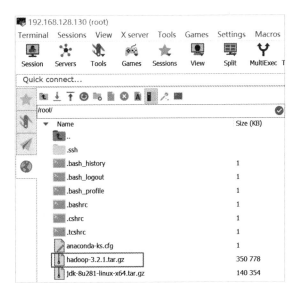

图 2-39　上传 Hadoop 软件包到 master 节点中

（2）在 master 节点中输入以下命令，解压缩 Hadoop 软件包并修改名称，操作过程与结果如图 2-40 所示。

```
# 解压缩 Hadoop 包到 "/usr/local" 目录下
[root@master ~]# tar -zxf ./hadoop-3.2.1.tar.gz -C /usr/local
# 进入 "local" 目录
[root@master ~]# cd /usr/local
# 查看 "local" 目录下的内容
[root@master local]# ls
# 修改名称 hadoop-3.2.1 为 hadoop
```

[root@master local]# mv hadoop-3.2.1/ hadoop
再次查看"local"目录下的内容，确认修改成功
[root@master local]# ls

```
[root@master ~]# tar -zxf ./hadoop-3.2.1.tar.gz -C /usr/local
[root@master ~]#
[root@master ~]# cd /usr/local
[root@master local]# ls
bin  etc  games  hadoop-3.2.1  include  jdk1.8.0_281  lib  lib64  libexec  sbin  share  src
[root@master local]# mv hadoop-3.2.1/ hadoop
[root@master local]# ls
bin  etc  games  hadoop  include  jdk1.8.0_281  lib  lib64  libexec  sbin  share  src
[root@master local]#
```

图 2-40　解压缩软件包并修改名称

（3）在 master 节点中输入以下命令，修改环境变量。

回到"root"目录
[root@master local]# cd
配置环境变量，添加 Hadoop 相关配置
[root@master ~]# vi .bash_profile

使用 Hadoop 解压缩后的路径设置 HADOOP_HOME 环境变量，添加 PATH 路径，如图 2-41 所示。

```
# .bash_profile

# Get the aliases and functions
if [ -f ~/.bashrc ]; then
        . ~/.bashrc
fi

# User specific environment and startup programs

PATH=$PATH:$HOME/bin
export JAVA_HOME=/usr/local/jdk1.8.0_281
export HADOOP_HOME=/usr/local/hadoop
PATH=$PATH:$JAVA_HOME/bin:$HADOOP_HOME/bin:$HADOOP_HOME/sbin
export PATH
~
```

图 2-41　在".bash_profile"文件中添加配置

（4）在 master 节点中输入以下命令，使环境变量生效，查看 Hadoop 版本，能正确显示 Hadoop 3.2.1 版本则说明 Hadoop 安装成功，结果如图 2-42 所示。

[root@master ~]# source .bash_profile
查看 Hadoop 版本
[root@master ~]# hadoop version

```
[root@master ~]# vi .bash_profile
[root@master ~]# source .bash_profile
[root@master ~]# hadoop version
Hadoop 3.2.1
Source code repository https://gitbox.apache.org/repos/asf/hadoop.git -r b3cbbb467e22ea829b3808f4b7b
01d07e0bf3842
Compiled by rohithsharmaks on 2019-09-10T15:56Z
Compiled with protoc 2.5.0
From source with checksum 776eaf9eee9c0ffc370bcbc1888737
This command was run using /usr/local/hadoop/share/hadoop/common/hadoop-common-3.2.1.jar
[root@master ~]#
```

图 2-42　查看 Hadoop 版本，验证安装是否成功

2. 修改 Hadoop 配置文件

1）查看 Hadoop 的所有配置文件

在 master 节点中输入以下命令，进入 Hadoop 配置文件所在目录，查看所有配置文件，结果如图 2-43 所示。

[root@master ~]# cd /usr/local/hadoop/etc/hadoop/
[root@master hadoop]# ls

```
[root@master ~]# cd /usr/local/hadoop/etc/hadoop/
[root@master hadoop]# ls
capacity-scheduler.xml              httpfs-log4j.properties       mapred-site.xml
configuration.xsl                   httpfs-signature.secret       shellprofile.d
container-executor.cfg              httpfs-site.xml               ssl-client.xml.example
core-site.xml                       kms-acls.xml                  ssl-server.xml.example
hadoop-env.cmd                      kms-env.sh                    user_ec_policies.xml.template
hadoop-env.sh                       kms-log4j.properties          workers
hadoop-metrics2.properties          kms-site.xml                  yarn-env.cmd
hadoop-policy.xml                   log4j.properties              yarn-env.sh
hadoop-user-functions.sh.example    mapred-env.cmd                yarnservice-log4j.properties
hdfs-site.xml                       mapred-env.sh                 yarn-site.xml
httpfs-env.sh                       mapred-queues.xml.template
[root@master hadoop]#
```

图 2-43 查看 Hadoop 的所有配置文件

在这些配置文件中，主要修改的文件有"hadoop-env.sh""core-site.xml""hdfs-site.xml""mapred-site.xml""yarn-site.xml""workers"。

2）配置"hadoop-env.sh"文件

"hadoop-env.sh"文件主要用于配置 Hadoop 的环境变量。在此文件的末尾添加 JAVA_HOME 的目录，定义 HDFS 和 YARN 的相关角色用户为 root 用户。

在 master 节点中输入以下命令，修改文件配置。

[root@master hadoop]# vi hadoop-env.sh

在"hadoop-env.sh"文件末尾增加以下内容，如图 2-44 所示，保存并退出编辑。

export JAVA_HOME=/usr/local/jdk1.8.0_281/
export HDFS_NAMENODE_USER=root
export HDFS_DATANODE_USER=root
export HDFS_SECONDARYNAMENODE_USER=root
export YARN_RESOURCEMANAGER_USER=root
export YARN_NODEMANAGER_USER=root

```
# To prevent accidents, shell commands be (superficially) locked
# to only allow certain users to execute certain subcommands.
# It uses the format of (command)_(subcommand)_USER.
#
# For example, to limit who can execute the namenode command,
# export HDFS_NAMENODE_USER=hdfs
export JAVA_HOME=/usr/local/jdk1.8.0_281/
export HDFS_NAMENODE_USER=root
export HDFS_DATANODE_USER=root
export HDFS_SECONDARYNAMENODE_USER=root
export YARN_RESOURCEMANAGER_USER=root
export YARN_NODEMANAGER_USER=root
-- INSERT --
```

图 2-44 修改"hadoop-env.sh"文件

3）配置"core-site.xml"文件

"core-site.xml"文件是 Hadoop 的核心配置项，如配置 HDFS 和 MapReduce 常用的 I/O 设置等。

在 master 节点中输入以下命令，修改文件配置。

[root@master hadoop]# vi core-site.xml

在该文件的 <configuration> 和 </configuration> 标签之间写入以下配置内容，结果如图 2-45 所示。

```
<configuration>
  <property>
    <!-- file 表示本地目录，Hadoop 的临时目录 -->
    <name>hadoop.tmp.dir</name>
    <value>file:/usr/local/hadoop/tmp</value>
  </property>
  <property>
    <!-- HDFS 的 NameNode 默认文件系统通信地址（主机＋端口） -->
    <name>fs.defaultFS</name>
    <value>hdfs://master:8020</value>
  </property>
</configuration>
```

图 2-45 修改"core-site.xml"文件

在以上配置信息中，必须要配置 hadoop.tmp.dir 参数。如果没有配置该参数，则系统会使用默认的目录作为其临时目录。该默认目录在每次系统重启时都会被删除，必须重新执行 Hadoop 文件系统格式化命令，否则运行 Hadoop 就会报错。

4）配置"hdfs-site.xml"文件

"hdfs-site.xml"文件主要配置以下内容。

（1）NameNode 与 secondaryNameNode 的访问地址。

（2）NameNode 与 DataNode 数据的存放路径。

（3）FSImage、Edits、Checkpoint 的存放位置。

（4）文件的副本数，即将一份文件保存为多少份。

（5）文件存储的 block 块大小为 128MB。

在 master 节点中输入以下命令，修改文件配置。

[root@master hadoop]# vi hdfs-site.xml

在该文件的 <configuration> 和 </configuration> 标签之间写入以下配置内容，结果如图 2-46 所示。

```
<configuration>
  <property>
    <!-- 指定 DataNode 文件的副本数，要小于节点数 -->
    <name>dfs.replication</name>
    <value>3</value>
  </property>
  <property>
    <!-- 配置 NameNode 的临时文件目录地址，记得在该位置新建"name"文件夹 -->
    <name>dfs.namenode.name.dir</name>
    <value>/usr/local/hadoop/hdfs/name</value>
  </property>
  <property>
    <!-- 配置 DataNode 的临时文件目录地址，记得在该位置新建"data"文件夹 -->
    <name>dfs.datanode.data.dir</name>
    <value>/usr/local/hadoop/hdfs/data</value>
  </property>
  <property>
    <!-- 是否开启 webhdfs 的 UI 监控，如果 true 则集群安全性会较差 -->
    <name>dfs.webhdfs.enabled</name>
    <value>true</value>
  </property>
</configuration>
```

```
Unless required by applicable law or agreed to in writing, software
distributed under the License is distributed on an "AS IS" BASIS,
WITHOUT WARRANTIES OR CONDITIONS OF ANY KIND, either express or implied.
See the License for the specific language governing permissions and
limitations under the License. See accompanying LICENSE file.
-->

<!-- Put site-specific property overrides in this file. -->

<configuration>
        <property>
                <!--指定DataNode文件的副本数，要小于节点数-->
                <name>dfs.replication</name>
                <value>3</value>
        </property>
        <property>
                <!--配置NameNode的临时文件目录地址，记得在该位置新建"name"文件夹-->
                <name>dfs.namenode.name.dir</name>
                <value>/usr/local/hadoop/hdfs/name</value>
        </property>
        <property>
                <!--配置DataNode的临时文件目录地址，记得在该位置新建"data"文件夹-->
                <name>dfs.datanode.data.dir</name>
                <value>/usr/local/hadoop/hdfs/data</value>
        </property>
        <property>
                <!--是否开启webhdfs的UI监控，如果true则集群安全性会较差-->
                <name>dfs.webhdfs.enabled</name>
                <value>true</value>
        </property>
</configuration>
-- INSERT --
```

图 2-46　修改 "hdfs-site.xml" 文件

5) 配置 "mapred-site.xml" 文件

"mapred-site.xml" 文件是 MapReduce 守护进程的配置项，包括作业历史服务器等。在 master 节点中输入以下命令，修改文件配置。

[root@master hadoop]# vi mapred-site.xml

在该文件的 <configuration> 和 </configuration> 标签之间写入以下配置内容，结果如图 2-47 所示。

```
<configuration>
  <property>
    <name>mapreduce.framework.name</name>
    <value>yarn</value>
    <description>指的是使用 YARN 运行 MapReduce 程序 </description>
  </property>
  <property>
    <name>mapreduce.jobhistory.address</name>
    <value>master:10020</value>
    <description>JobHistory 用来记录已经 finished（结束）的 MapReduce 运行日志，日志信息存放于 HDFS 目录中，在默认情况下没有开启此功能 </description>
  </property>
  <property>
    <name>mapreduce.jobhistory.webapp.address</name>
    <value>master:19888</value>
    <description>定义 MapReduce 运行日志的 Web 访问地址和端口 </description>
  </property>
```

```xml
<property>
    <name>yarn.app.mapreduce.am.env</name>
    <value>HADOOP_MAPRED_HOME=/usr/local/hadoop</value>
</property>
<property>
    <name>mapreduce.map.env</name>
    <value>HADOOP_MAPRED_HOME=/usr/local/hadoop</value>
</property>
<property>
    <name>mapreduce.reduce.env</name>
    <value>HADOOP_MAPRED_HOME=/usr/local/hadoop</value>
</property>
</configuration>
```

```
<!-- Put site-specific property overrides in this file. -->
<configuration>
    <property>
        <name>mapreduce.framework.name</name>
        <value>yarn</value>
        <description>指的是使用YARN运行MapReduce程序</description>
    </property>
    <property>
        <name>mapreduce.jobhistory.address</name>
        <value>master:10020</value>
        <description>JobHistory用来记录已经finished（结束）的MapReduce运行日志，日志信息存放于HDFS目录中，在默认情况下没有开启此功能</description>
    </property>
    <property>
        <name>mapreduce.jobhistory.webapp.address</name>
        <value>master:19888</value>
        <description>MapReduce运行日志的Web访问地址</description>
    </property>
    <property>
        <name>yarn.app.mapreduce.am.env</name>
        <value>HADOOP_MAPRED_HOME=/usr/local/hadoop</value>
    </property>
    <property>
        <name>mapreduce.map.env</name>
        <value>HADOOP_MAPRED_HOME=/usr/local/hadoop</value>
    </property>
    <property>
        <name>mapreduce.reduce.env</name>
        <value>HADOOP_MAPRED_HOME=/usr/local/hadoop</value>
    </property>
</configuration>
```

图 2-47　修改"mapred-site.xml"文件

6）配置"yarn-site.xml"文件

该文件是 YARN 守护进程的配置项，包括资源管理器、Web 应用代理服务器和节点管理器等。

在 master 节点中输入以下命令，修改文件配置。

[root@master hadoop]# vi yarn-site.xml

在该文件的 <configuration> 和 </configuration> 标签之间写入以下配置内容，结果如图 2-48 所示。

<!-- 以下涉及的简写有 RM：ResourceManager；AM：ApplicationMaster；NM：NodeManager -->

```xml
<configuration>
  <property>
    <name>yarn.resourcemanager.address</name>
    <value>master:8032</value>
    <description>RM 对客户端暴露的地址，客户端通过该地址向 RM 提交应用程序等</description>
  </property>
  <property>
    <name>yarn.resourcemanager.scheduler.address</name>
    <value>master:8030</value>
    <description>RM 对 AM 暴露的地址，AM 通过该地址向 RM 申请资源，释放资源等</description>
  </property>
  <property>
    <name>yarn.resourcemanager.resource-tracker.address</name>
    <value>master:8031</value>
    <description>RM 对 NM 暴露的地址，NM 通过该地址向 RM 汇报心跳，领取任务等</description>
  </property>
  <property>
    <name>yarn.resourcemanager.admin.address</name>
    <value>master:8033</value>
    <description>管理员可以通过该地址向 RM 发送管理命令等</description>
  </property>
  <property>
    <name>yarn.resourcemanager.webapp.address</name>
    <value>master:8088</value>
    <description>RM 对外暴露的 web http 地址，用户可以通过该地址在浏览器中查看集群信息</description>
  </property>
  <property>
    <name>yarn.nodemanager.aux-services</name>
    <value>mapreduce_shuffle</value>
    <description>NM 上运行的附属服务，需配置成 mapreduce_shuffle，才可运行 MapReduce 程序，默认值为 ""</description>
  </property>
  <property>
    <name>yarn.resourcemanager.hostname</name>
    <value>master</value>
    <description>指定 resourcemanager 的 hostname 为 master</description>
  </property>
</configuration>
```

```xml
See the License for the specific language governing permissions and
limitations under the License. See accompanying LICENSE file.
-->
<configuration>
    <property>
        <name>yarn.resourcemanager.address</name>
        <value>master:8032</value>
        <description>RM对客户端暴露的地址,客户端通过该地址向RM提交应用程序等</description>
    </property>
    <property>
        <name>yarn.resourcemanager.scheduler.address</name>
        <value>master:8030</value>
        <description>RM对AM暴露的地址,AM通过该地址向RM申请资源,释放资源等</description>
    </property>
    <property>
        <name>yarn.resourcemanager.resource-tracker.address</name>
        <value>master:8031</value>
        <description>RM对NM暴露的地址,NM通过该地址向RM汇报心跳,领取任务等</description>
    </property>
    <property>
        <name>yarn.resourcemanager.admin.address</name>
        <value>master:8033</value>
        <description>管理员可以通过该地址向RM发送管理命令等</description>
    </property>
    <property>
        <name>yarn.resourcemanager.webapp.address</name>
        <value>master:8088</value>
        <description>RM对外暴露的web http地址,用户可以通过该地址在浏览器中查看集群信息</description>
    </property>
    <property>
        <name>yarn.nodemanager.aux-services</name>
        <value>mapreduce_shuffle</value>
        <description>NM上运行的附属服务。需配置成mapreduce_shuffle,才可运行MapReduce程序,默认值为""</description>
    </property>
    <property>
        <name>yarn.resourcemanager.hostname</name>
        <value>master</value>
        <description>指定resourcemanager的hostname为master</description>
    </property>
</configuration>
```

图 2-48　修改"yarn-site.xml"文件

7)配置"workers"文件

"workers"文件用于记录所有数据节点的主机名或 IP 地址。

在 master 节点中输入以下命令,修改文件配置。

[root@master hadoop]# vi workers

配置内容如图 2-49 所示。

```
localhost
slaver01
slaver02
~
```

图 2-49　在"workers"文件中添加配置

3. 分发 Hadoop 相关文件和环境变量文件

在上述文件全部配置完成之后,需要将 master 节点中的"/usr/local/hadoop/"文件夹和"root"目录下的".bash_profile"环境变量文件复制到 slaver01 和 slaver02 节点中,实现配置文件的同步。

(1)分发文件给 slaver01。

[root@master hadoop]# cd
分发 Hadoop 相关文件
[root@master ~]# scp -r /usr/local/hadoop/ root@slaver01:/usr/local
分发环境变量文件
[root@master ~]# scp .bash_profile root@slaver01:/root

进入 slaver01 确认分发结果
[root@slaver01 ~]# source .bash_profile
[root@slaver01 ~]# hadoop version

查看 Hadoop 版本为 Hadoop 3.2.1，如图 2-50 所示，则表示安装配置成功。

```
[root@slaver01 ~]# source .bash_profile
[root@slaver01 ~]# hadoop version
Hadoop 3.2.1
Source code repository https://gitbox.apache.org/repos/asf/hadoop.git -r b3cbbb467e22ea829b3808f4b7b
01d07e0bf3842
Compiled by rohithsharmaks on 2019-09-10T15:56Z
Compiled with protoc 2.5.0
From source with checksum 776eaf9eee9c0ffc370bcbc1888737
This command was run using /usr/local/hadoop/share/hadoop/common/hadoop-common-3.2.1.jar
[root@slaver01 ~]#
```

图 2-50　在 slaver01 节点中确认分发结果

（2）分发文件给 slaver02。

分发 Hadoop 相关文件
[root@master ~]# scp -r /usr/local/hadoop/ root@slaver02:/usr/local
分发环境变量文件
[root@master ~]# scp .bash_profile root@slaver02:/root
进入 slaver02 确认分发结果
[root@slaver02 ~]# source .bash_profile
[root@slaver02 ~]# hadoop version

分发过程与结果如图 2-51 所示，安装配置成功。

```
[root@slaver02 ~]# source .bash_profile
[root@slaver02 ~]# hadoop version
Hadoop 3.2.1
Source code repository https://gitbox.apache.org/repos/asf/hadoop.git -r b3cbbb467e22ea829b3808f4b7b
01d07e0bf3842
Compiled by rohithsharmaks on 2019-09-10T15:56Z
Compiled with protoc 2.5.0
From source with checksum 776eaf9eee9c0ffc370bcbc1888737
This command was run using /usr/local/hadoop/share/hadoop/common/hadoop-common-3.2.1.jar
[root@slaver02 ~]#
```

图 2-51　在 slaver02 节点中确认分发结果

4. 格式化 HDFS

Hadoop 生态中的文件系统类似一块磁盘，在初次使用时需要格式化，让存储空间明白该按什么方式组织存储数据。

格式化 HDFS，此操作仅需执行一次。在 master 节点中输入以下命令，格式化结果如图 2-52 所示。

[root@master ~]# hdfs namenode -format

```
2022-10-19 05:04:15,855 INFO namenode.FSImage: Allocated new BlockPoolId: BP-1519610361-192.1
68.128.130-1666170255833
2022-10-19 05:04:15,888 INFO common.Storage: Storage directory /usr/local/hadoop/hdfs/name ha
s been successfully formatted.
2022-10-19 05:04:16,000 INFO namenode.FSImageFormatProtobuf: Saving image file /usr/local/had
oop/hdfs/name/current/fsimage.ckpt_0000000000000000000 using no compression
2022-10-19 05:04:16,189 INFO namenode.FSImageFormatProtobuf: Image file /usr/local/hadoop/hdf
s/name/current/fsimage.ckpt_0000000000000000000 of size 399 bytes saved in 0 seconds .
2022-10-19 05:04:16,204 INFO namenode.NNStorageRetentionManager: Going to retain 1 images wit
h txid >= 0
2022-10-19 05:04:16,214 INFO namenode.FSImage: FSImageSaver clean checkpoint: txid=0 when mee
t shutdown.
2022-10-19 05:04:16,215 INFO namenode.NameNode: SHUTDOWN_MSG:
/************************************************************
SHUTDOWN_MSG: Shutting down NameNode at master/192.168.128.130
************************************************************/
[root@master ~]#
```

图 2-52　格式化结果

2.1.3　启动 Hadoop 集群

1. 启动 Hadoop 集群

启动 Hadoop 集群只需要在 master 节点中输入以下命令，结果如图 2-53 所示。

[root@master ~]# start-all.sh

```
[root@master ~]# start-all.sh
Starting namenodes on [master]
Last login: Wed Oct 19 02:20:57 EDT 2022 from localhost on pts/2
Starting datanodes
Last login: Wed Oct 19 05:12:50 EDT 2022 on pts/1
slaver02: WARNING: /usr/local/hadoop/logs does not exist. Creating.
slaver01: WARNING: /usr/local/hadoop/logs does not exist. Creating.
Starting secondary namenodes [master]
Last login: Wed Oct 19 05:12:54 EDT 2022 on pts/1
Starting resourcemanager
Last login: Wed Oct 19 05:13:08 EDT 2022 on pts/1
Starting nodemanagers
Last login: Wed Oct 19 05:13:22 EDT 2022 on pts/1
```

图 2-53　启动 Hadoop 集群

2. 确认 Hadoop 集群启动是否成功

查看各节点的服务进程，如图 2-54、图 2-55、图 2-56 所示，在 master 节点中可以看到 NameNode、DataNode、ResourceManager、SecondaryNameNode、NodeManager 这 5 个服务进程；在 slaver01 和 slaver02 节点中可以看到 DataNode、NodeManager 这两个服务进程，则表示 Hadoop 集群启动成功。

```
[root@master ~]# jps
9923 NameNode
10276 SecondaryNameNode
10660 NodeManager
10982 Jps
10056 DataNode
10522 ResourceManager
[root@master ~]#
```

图 2-54　master 节点确认集群启动成功

```
[root@slaver01 ~]# jps
9889 NodeManager
9783 DataNode
9982 Jps
[root@slaver01 ~]#
```

```
[root@slaver02 ~]# jps
9765 DataNode
9963 Jps
9871 NodeManager
[root@slaver02 ~]#
```

图 2-55　slaver01 节点确认集群启动成功　　　　图 2-56　slaver02 节点确认集群启动成功

3. 查看网页并确认安装部署成功

1）打开 YARN 页面

打开 Google Chrome，输入地址"192.168.128.130:8088"，可以打开 YARN 页面，如图 2-57 所示。

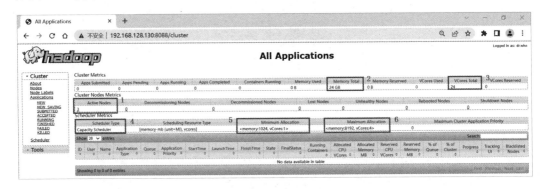

图 2-57　YARN 页面

在 YARN 页面中可以获得许多信息。

（1）Active Nodes：表示 YARN 集群管理节点的个数，即 NodeManager 的个数，本集群有 3 个 NodeManager。

（2）Memory Total：表示 YARN 集群管理内存的总和，等于所有 NodeManager 管理的内存之和。

（3）VCores Total：表示 YARN 集群管理 CPU 虚拟核心的总和，等于所有 NodeManager 管理的虚拟核心之和。

（4）Scheduler Type：表示资源分配的类型。

（5）Minimum Allocation：最小分配资源，即当一个任务向 YARN 申请资源时，YARN 至少会分配 <memory:1024,vCores:1> 资源给这个任务。分配的最小内存可以由 "yarn-site.xml" 配置文件中的 yarn.scheduler.minimum-allocation-mb（默认值为 1024，单位为 MB）来控制，最小核心数可以由 "yarn-site.xml" 配置文件中的 yarn.scheduler.minimum-allocation-vcores（默认值为 1）来控制。

（6）Maximum Allocation：最大分配资源，即当一个任务向 YARN 申请资源时，YARN 最多会分配 <memory:1630,vCores:2> 资源给这个任务。分配的最大内存由 "yarn-site.xml" 配置文件中的 yarn.scheduler.maximum-allocation-mb（默认值为 8192，单位为 MB）来控制，最大核心数可以由 "yarn-site.xml" 配置文件中的 yarn.scheduler.maximum-allocation-vcores（默认值为 32）来控制，这两个值不能大于集群管理的资源总和。

2）打开 HDFS 页面

打开 Google Chrome，输入地址"192.168.128.130:9870"，可以打开 HDFS 页面，如图 2-58 所示。

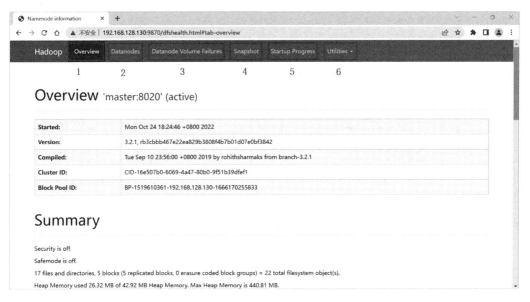

图 2-58　HDFS 页面

在 HDFS 页面中可以获得如下信息。

（1）Overview：集群概述，介绍了集群的基本情况，包括版本、集群 ID、安全模式、数据节点文件系统的容量和使用率、数据存储的路径等基本信息。

（2）Datanodes：数据节点，包括数据节点使用率柱状图、每个数据节点的磁盘使用率、运行中的节点、进入维护的节点列表、退役的节点列表等信息。

（3）Datanode Volume Failures：数据节点卷故障。

（4）Snapshot：快照，包括快照摘要、快照目录列表、已创建的快照目录等信息。

（5）Startup Progress：启动进度，包括集群启动时加载的 fsimage 和 edits 信息。

（6）Utilities：包括集群 DFS 存储系统的可视化浏览、各个集群组件的日志等信息。

3）查看 IP 地址和端口地址

除了使用 Web 页面查看 Hadoop 安装部署情况，还可以使用"netstat-ntpl"命令查看 IP 地址和端口地址是否正常，各端口启动情况如图 2-59 所示。

在 master 节点中输入以下命令。

[root@master ~]# netstat -ntpl

如果没有这个命令，则需要先安装 net-tools 软件。net-tools 包含 ifconfig、route、arp 和 netstat 等命令行工具，用于管理和排查各种网络配置，其安装命令是"yum install net-tools"。

```
[root@master ~]# netstat -ntpl
Active Internet connections (only servers)
Proto Recv-Q Send-Q Local Address           Foreign Address         State       PID/Program name
tcp        0      0 192.168.128.130:8030    0.0.0.0:*               LISTEN      2140/java
tcp        0      0 192.168.128.130:8031    0.0.0.0:*               LISTEN      2140/java
tcp        0      0 192.168.128.130:8032    0.0.0.0:*               LISTEN      2140/java
tcp        0      0 192.168.128.130:8033    0.0.0.0:*               LISTEN      2140/java
tcp        0      0 0.0.0.0:8040            0.0.0.0:*               LISTEN      2288/java
tcp        0      0 0.0.0.0:9864            0.0.0.0:*               LISTEN      1658/java
tcp        0      0 0.0.0.0:8042            0.0.0.0:*               LISTEN      2288/java
tcp        0      0 0.0.0.0:9866            0.0.0.0:*               LISTEN      1658/java
tcp        0      0 0.0.0.0:9867            0.0.0.0:*               LISTEN      1658/java
tcp        0      0 0.0.0.0:9868            0.0.0.0:*               LISTEN      1870/java
tcp        0      0 0.0.0.0:9870            0.0.0.0:*               LISTEN      1520/java
tcp        0      0 0.0.0.0:46002           0.0.0.0:*               LISTEN      2288/java
tcp        0      0 192.168.128.130:8020    0.0.0.0:*               LISTEN      1520/java
tcp        0      0 0.0.0.0:22              0.0.0.0:*               LISTEN      899/sshd
tcp        0      0 127.0.0.1:35863         0.0.0.0:*               LISTEN      1658/java
tcp        0      0 192.168.128.130:8088    0.0.0.0:*               LISTEN      2140/java
tcp        0      0 127.0.0.1:25            0.0.0.0:*               LISTEN      1147/master
tcp        0      0 0.0.0.0:13562           0.0.0.0:*               LISTEN      2288/java
tcp6       0      0 :::22                   :::*                    LISTEN      899/sshd
tcp6       0      0 ::1:25                  :::*                    LISTEN      1147/master
[root@master ~]#
```

图 2-59　查看各端口启动情况

2.1.4　运行经典案例 wordcount

通过经典案例 wordcount，体验 Hadoop 运行 MapReduce 计算。

1. 新建一个文本文件

在本地"root"目录下创建一个文本文件"wordcount"，结果如图 2-60 所示。

[root@master ~]# vi wordcount.txt
[root@master ~]# cat wordcount.txt

```
[root@master ~]# vi wordcount.txt
[root@master ~]# cat wordcount.txt
hello world
hello hadoop
hello spark
hello scala
[root@master ~]#
```

图 2-60　新建"wordcount"文本文件

2. 上传文本文件到 HDFS 中

\# 在 HDFS 中新建一个"input"目录
[root@master ~]# hdfs dfs -mkdir /input
\# 将"wordcount.txt"文件上传到"input"目录中
[root@master ~]# hdfs dfs -put wordcount.txt /input
\# 查看文件是否上传成功
[root@master ~]# hdfs dfs -ls /input

操作过程与结果如图 2-61 所示。

```
[root@master ~]# hdfs dfs -mkdir /input
[root@master ~]# hdfs dfs -put wordcount.txt /input
2022-10-19 06:01:10,051 INFO sasl.SaslDataTransferClient: SASL encryption trust check: localHostTrus
ted = false, remoteHostTrusted = false
[root@master ~]# hdfs dfs -ls /input
Found 1 items
-rw-r--r--   3 root supergroup         49 2022-10-19 06:01 /input/wordcount.txt
[root@master ~]#
```

图 2-61　上传文本文件到 HDFS 中并查看

3. 运行 wordcount 并查看结果

\# 进入 Hadoop 自带的 "hadoop-mapreduce-examples-3.2.1.jar" 包目录

[root@master ~]# cd /usr/local/hadoop/share/hadoop/mapreduce

\# 运行 wordcount 并将结果输出到 "output" 目录中

[root@master mapreduce]# hadoop jar hadoop-mapreduce-examples-3.2.1.jar wordcount /input /output

第二条执行命令中各参数的含义如下。

hadoop："$HADOOP_HOME/bin" 下的 shell 脚本名。

jar：hadoop 脚本需要的 command 参数。

hadoop-mapreduce-examples-3.2.1.jar：要执行的 jar 包在本地文件系统中的完整路径。

wordcount：main() 方法所在的类。

/input：HDFS 的路径，指示输入数据来源。

/output：HDFS 的路径，指示输出数据路径。

结果显示 successfully，则表示运行成功，如图 2-62 所示。

```
2022-10-19 06:04:16,670 INFO mapreduce.Job:  map 100% reduce 100%
2022-10-19 06:04:16,706 INFO mapreduce.Job: Job job_1666173575901_0001 complet
ed successfully
2022-10-19 06:04:16,914 INFO mapreduce.Job: Counters: 54
        File System Counters
                FILE: Number of bytes read=67
                FILE: Number of bytes written=452065
                FILE: Number of read operations=0
                FILE: Number of large read operations=0
                FILE: Number of write operations=0
```

图 2-62　运行结果

查看计算结果，在 master 节点中输入以下命令。

[root@master mapreduce]# hdfs dfs -cat /output/part-r-00000

```
[root@master mapreduce]# hdfs dfs -cat /output/part-r-00000
2022-10-19 06:06:46,952 INFO sasl.SaslDataTransferClient: SASL encryption trus
t check: localHostTrusted = false, remoteHostTrusted = false
hadoop  1
hello   4
scala   1
spark   1
world   1
[root@master mapreduce]#
```

图 2-63　查看计算结果

从计算结果中可以看到计算获得的各个单词的个数。

任务 2.2 Spark 集群部署与使用

情境导入

Spark SQL 是 Spark 的组件之一，要利用 Spark SQL 工具进行数据分析，需要先安装部署 Spark。Spark 可以搭建单机环境、伪分布式环境、完全分布式环境，并根据需求进行安装。在一般的实际应用中，Spark 会以集群的方式运行，从而实现分布式计算。本任务将逐步进行 Spark 在完全分布式环境中的安装部署，以及详细的文件配置部署。

学习目标和要求

知识与技能目标
1. 掌握 Spark 集群安装配置方法。
2. 掌握启动、关闭 Spark 集群的方法。
3. 能使用 Spark-shell 进行简单的编程测试。

素质目标
1. 注重工作效率。
2. 具有解决问题的灵活应变能力。

2.2.1 安装 Spark

Spark 安装

本书将在 master、slaver01、slaver02 这 3 个节点中安装部署 Spark。为了提高安装部署效率，可以先在 master 节点中进行安装，再将相关配置文件复制并发送到另外两个节点中。本书安装的 Spark 版本是 Spark-3.1.2。

1. 上传、解压缩、重命名软件包

使用 MobaXterm 工具，将软件包上传到 master 节点的"root"目录中，将其解压缩到"/usr/local"目录中。

在 master 节点中输入以下命令，对软件包进行解压缩和重命名，操作过程与结果如图 2-64 所示。

```
[root@master ~]# tar -zxf spark-3.1.2-bin-hadoop3.2.tgz -C /usr/local/
[root@master ~]# mv /usr/local/spark-3.1.2-bin-hadoop3.2/ /usr/local/spark
[root@master ~]# ls /usr/local/
```

```
[root@master ~]# tar -zxf spark-3.1.2-bin-hadoop3.2.tgz -C /usr/local/
[root@master ~]# mv /usr/local/spark-3.1.2-bin-hadoop3.2/ /usr/local/spark
[root@master ~]# ls /usr/local/
bin   games    include         lib     libexec   share   src
etc   hadoop   jdk1.8.0_281    lib64   sbin      spark
[root@master ~]#
```

图 2-64　解压缩 Spark 软件包并重命名

2．修改环境变量文件并使其生效

（1）在 master 节点中输入以下命令，在 ".bash_profile" 文件中添加配置之后，查看此文件，操作结果如图 2-65 所示。

```
// 修改环境变量
[root@master ~]# vi .bash_profile
// 查看修改情况
[root@master ~]# cat .bash_profile
# .bash_profile

# Get the aliases and functions
if [ -f ~/.bashrc ]; then
    . ~/.bashrc
fi

# User specific environment and startup programs

PATH=$PATH:$HOME/bin
export JAVA_HOME=/usr/local/jdk1.8.0_281
export HADOOP_HOME=/usr/local/hadoop
export SPARK_HOME=/usr/local/spark
PATH=$PATH:$JAVA_HOME/bin:$HADOOP_HOME/bin:$HADOOP_HOME/sbin:$SPARK_HOME/bin:$SPARK_HOME/sbin
export PATH
```

```
[root@master ~]# vi .bash_profile
[root@master ~]# cat .bash_profile
# .bash_profile

# Get the aliases and functions
if [ -f ~/.bashrc ]; then
    . ~/.bashrc
fi

# User specific environment and startup programs

PATH=$PATH:$HOME/bin
export JAVA_HOME=/usr/local/jdk1.8.0_281
export HADOOP_HOME=/usr/local/hadoop
export SPARK_HOME=/usr/local/spark
PATH=$PATH:$JAVA_HOME/bin:$HADOOP_HOME/bin:$HADOOP_HOME/sbin:$SPARK_HOME/bin:$SPARK_HOME/sbin
export PATH
[root@master ~]#
```

图 2-65　修改环境变量文件

（2）使环境变量文件生效。

在 master 节点中输入以下命令，使环境变量文件生效。

[root@master ~]# source .bash_profile

在输入"spark"之后按两次"Tab"键进行补全操作，若能正常补全，则说明安装成功，结果如图 2-66 所示。

[root@master ~]# spark

```
[root@master ~]# source .bash_profile
[root@master ~]# spark
spark-class       spark-daemon.sh    spark-shell
spark-class2.cmd  spark-daemons.sh   spark-sql
spark-config.sh   sparkR             spark-submit
[root@master ~]# spark
```

图 2-66　使环境变量文件生效

3. 修改 Spark 配置文件

1）配置"spark-env.sh"文件

（1）在 master 节点中输入以下命令。

```
# 进入 Spark 配置文件目录
[root@master ~]# cd /usr/local/spark/conf
# 复制"spark-env.sh.template"模板文件，重命名为 spark-env.sh
[root@master conf]# cp spark-env.sh.template spark-env.sh
# 使用 vi 编辑器对"spark-env.sh"文件进行修改
[root@master conf]# vi spark-env.sh
```

在此文件末尾添加图 2-67 所示的内容。

```
export JAVA_HOME=/usr/local/jdk1.8.0_281
export HADOOP_HOME=/usr/local/hadoop
export HADOOP_CONF_DIR=$HADOOP_HOME/etc/hadoop
export SPARK_MASTER_HOST=master
export SPARK_LOCAL_DIRS=/usr/local/spark
export SPARK_DRIVER_MEMORY=1g       #内存
export SPARK_WORKER_CORES=1         #cpus核心数
```

图 2-67　"spark-env.sh"文件配置内容

（2）在"spark-env.sh"文件中添加的配置内容说明如下。

JAVA_HOME：Java 安装目录。

HADOOP_HOME：Hadoop 安装目录。

HADOOP_CONF_DIR：Hadoop 集群的配置文件的目录。

SPARK_MASTER_HOST：Spark 集群的 master 节点主机名。

SPARK_LOCAL_DIRS：Spark 集群的安装目录。

SPARK_DRIVER_MEMORY：每个 worker 节点能够分配给 exectors 的最大内存。

SPARK_WORKER_CORES：每个 worker 节点所占有的 CPU 核数目。

2）配置"workers"文件

"workers"文件用于记录所有数据节点的主机名，在 master 节点中输入以下命令。

复制 "workers.template" 模板文件为 "workers"
[root@master conf]# cp workers.template workers
[root@master conf]# vi workers

在"workers"文件末尾添加图 2-68 所示的内容。

```
# A Spark Worker will be started on each of the machines listed below.
localhost
slaver01
slaver02
~
```

图 2-68　"workers"文件配置内容

4. 重命名启动命令

复制并修改 Spark 启动脚本，避免和 Hadoop 的启动脚本冲突。在 master 节点中输入以下命令，操作过程与结果如图 2-69 所示。

进入 Spark 启动脚本目录
[root@master conf]# cd ../sbin/
查看、确认目录正确
[root@master sbin]# pwd
/usr/local/spark/sbin
对 Spark 启动脚本进行重命名
[root@master sbin]# cp start-all.sh start-spark-all.sh
[root@master sbin]# cp stop-all.sh stop-spark-all.sh
[root@master sbin]# ls

```
[root@master conf]# cd ../sbin/
[root@master sbin]# pwd
/usr/local/spark/sbin
[root@master sbin]# cp start-all.sh start-spark-all.sh
[root@master sbin]# cp stop-all.sh stop-spark-all.sh
[root@master sbin]# ls
decommission-slave.sh         start-worker.sh
decommission-worker.sh        start-workers.sh
slaves.sh                     stop-all.sh
spark-config.sh               stop-history-server.sh
spark-daemon.sh               stop-master.sh
spark-daemons.sh              stop-mesos-dispatcher.sh
start-all.sh                  stop-mesos-shuffle-service.sh
start-history-server.sh       stop-slave.sh
start-master.sh               stop-slaves.sh
start-mesos-dispatcher.sh     stop-spark-all.sh
start-mesos-shuffle-service.sh stop-thriftserver.sh
start-slave.sh                stop-worker.sh
start-slaves.sh               stop-workers.sh
start-spark-all.sh            workers.sh
start-thriftserver.sh
[root@master sbin]#
```

图 2-69　复制并修改 Spark 启动脚本

5. 分发"spark"文件

（1）分发"spark"文件到 slaver01 节点中。

在 master 节点中输入以下命令。

[root@master ~]# scp -r /usr/local/spark/ root@slaver01:/usr/local/
[root@master ~]# scp .bash_profile root@slaver01:/root/

使 slaver01 节点中的环境变量生效，并验证分发是否成功，结果如图 2-70 所示。

[root@slaver01 ~]# source .bash_profile
[root@slaver01 ~]# spark

```
[root@slaver01 ~]# source .bash_profile
[root@slaver01 ~]# spark
spark-class          spark-daemon.sh      spark-shell
spark-class2.cmd     spark-daemons.sh     spark-sql
spark-config.sh      sparkR               spark-submit
```

图 2-70　在 slaver01 节点中验证分发结果

（2）分发"spark"文件到 slaver02 节点中。

在 master 节点中输入以下命令。

[root@master ~]# scp -r /usr/local/spark/ root@slaver02:/usr/local/
[root@master ~]# scp .bash_profile root@slaver02:/root/

使 slaver02 节点中的环境变量生效，并验证是否分发成功，结果如图 2-71 所示。

[root@slaver02 ~]# source .bash_profile
[root@slaver02 ~]# spark

```
[root@slaver02 ~]# source .bash_profile
[root@slaver02 ~]# spark
spark-class          spark-daemon.sh      spark-shell
spark-class2.cmd     spark-daemons.sh     spark-sql
spark-config.sh      sparkR               spark-submit
```

图 2-71　在 slaver02 节点中验证分发结果

2.2.2　启动 Spark

1. 启动 Spark

在 master 节点中输入以下命令，启动 Spark，结果如图 2-72 所示。

[root@master ~]# start-spark-all.sh

```
[root@master ~]# start-spark-all.sh
starting org.apache.spark.deploy.master.Master, logging to /usr/local/spark/log
s/spark-root-org.apache.spark.deploy.master.Master-1-master.out
slaver02: starting org.apache.spark.deploy.worker.Worker, logging to /usr/local
/spark/logs/spark-root-org.apache.spark.deploy.worker.Worker-1-slaver02.out
slaver01: starting org.apache.spark.deploy.worker.Worker, logging to /usr/local
/spark/logs/spark-root-org.apache.spark.deploy.worker.Worker-1-slaver01.out
localhost: starting org.apache.spark.deploy.worker.Worker, logging to /usr/loca
l/spark/logs/spark-root-org.apache.spark.deploy.worker.Worker-1-master.out
```

图 2-72　启动 Spark

2. 查看各节点服务进程的状态（此处已经启动 Hadoop 集群）

（1）在 master 节点中输入以下命令。

```
[root@master ~]# jps
```

由图 2-73 可以看到在启动 Spark 之后，master 节点对应的 Worker、Master 服务进程也被启动。

```
[root@master ~]# jps
3664 Worker
2274 NodeManager
3730 Jps
1524 NameNode
1860 SecondaryNameNode
2133 ResourceManager
1660 DataNode
3612 Master
[root@master ~]#
```

图 2-73　查看 master 节点服务进程的状态

（2）在 slaver01 节点中输入以下命令。

```
[root@slaver01 ~]# jps
```

由图 2-74 可见，在 slaver01 节点中增加了 Worker 服务进程。

```
[root@slaver01 ~]# jps
2058 Jps
1499 NodeManager
2013 Worker
1391 DataNode
[root@slaver01 ~]#
```

图 2-74　查看 slaver01 节点服务进程的状态

（3）在 slaver02 节点中输入以下命令。

```
[root@slaver02 ~]# jps
```

由图 2-75 可见，在 slaver02 节点中增加了 Worker 服务进程。

```
[root@slaver02 ~]# jps
2211 Worker
1497 NodeManager
2268 Jps
1389 DataNode
[root@slaver02 ~]#
```

图 2-75　查看 slaver02 节点服务进程的状态

3. Web 查看 Spark 主页的情况

在 Google Chrome 中输入地址“http://192.168.128.130:8080/”，查看 Spark 主页的情况，如图 2-76 所示。

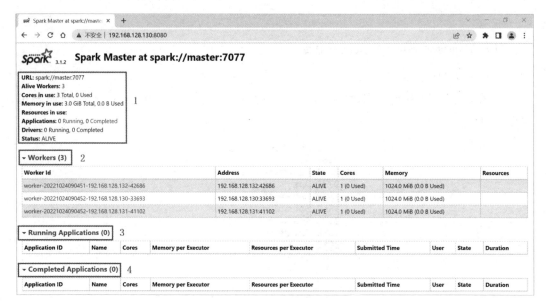

图 2-76　Web 查看 Spark 主页的情况

由图 2-76 可知 Spark 集群安装完成。在 Spark 主页中可以获得以下信息。

（1）任务信息统计：显示集群主节点的 URL、活着的节点、内核资源和存储的使用情况、运行中和已完成的任务情况等信息。

（2）Workers：显示运行中的节点相关信息，包括 Worker Id、Address、State、Cores、Memory、Resources 等信息。

（3）Running Applications 和 Completed Applications：显示运行中和已完成的任务情况，包括 Application ID、Name、Cores、Memory per Executor、Resources per Executor、Submitted Time、User、State、Duration 等信息。

2.2.3　Spark 集群测试

Spark 集群测试

1. 使用 spark-submit 工具提交 Spark 作业

对于数据的批处理，通常采用编写程序、打 .jar 包的方式提交给集群执行，这需要使用 Spark 自带的 spark-submit 工具。本案例将使用 spark-submit 工具提交 Spark 自带的样例程序——SparkPi，并计算 pi 值。

（1）在 master 节点中输入以下命令，结果如图 2-77 所示。

```
spark-submit \
--class org.apache.spark.examples.SparkPi \
--master spark://master:7077 \
--driver-memory 512M \
--executor-memory 512M \
--total-executor-cores 1 \
$SPARK_HOME/examples/jars/spark-examples_2.12-3.1.2.jar \
```

10

```
[root@master ~]# spark-submit \
> --class org.apache.spark.examples.SparkPi \
> --master spark://master:7077 \
> --driver-memory 512M \
> --executor-memory 512M \
> --total-executor-cores 1 \
> $SPARK_HOME/examples/jars/spark-examples_2.12-3.1.2.jar \
> 10
2022-10-19 10:51:01,588 INFO scheduler.TaskSchedulerImpl: Killing all running t
asks in stage 0: Stage finished
2022-10-19 10:51:01,595 INFO scheduler.DAGScheduler: Job 0 finished: reduce at
SparkPi.scala:38, took 7.815384 s
Pi is roughly 3.140895140895141
2022-10-19 10:51:01,624 INFO server.AbstractConnector: Stopped Spark@60d84f61{H
TTP/1.1, (http/1.1)}{0.0.0.0:4041}
2022-10-19 10:51:01,628 INFO ui.SparkUI: Stopped Spark web UI at http://master:
4041
```

图 2-77　使用 spark-submit 工具提交 Spark 作业

（2）spark-submit 提交任务的参数说明。

--class：应用程序的主类，仅针对 Java 或 Scala 应用。

--master：master 的地址，表示提交任务的执行位置，如 "spark://host:port,yarn,local"。

--driver-memory：driver 进程使用的内存大小，以字节为单位。可以指定不同的后缀，如 "512m" 或 "15g"，默认是 1G。

--executor-memory：executor 使用的内存大小，以字节为单位。可以指定不同的后缀，如 "512m" 或 "15g"，默认是 1G。

--total-executor-cores：所有 executor 的总核数，仅在 mesos 或者 standalone 下使用。

2. 使用 spark-sql

我们可以在 Spark SQL 中使用 SQL 命令进行数据的相关查询处理。

（1）启动 spark-sql。

在 master 节点中输入以下命令，启动 spark-sql，结果如图 2-78 所示。

[root@master ~]# spark-sql

```
[root@master ~]# spark-sql
2022-11-13 06:32:50,907 WARN util.NativeCodeLoader: Unable to load native-hadoop lib
rary for your platform... using builtin-java classes where applicable
Setting default log level to "WARN".
To adjust logging level use sc.setLogLevel(newLevel). For SparkR, use setLogLevel(ne
wLevel).
2022-11-13 06:32:58,250 WARN conf.HiveConf: HiveConf of name hive.stats.jdbc.timeout
 does not exist
2022-11-13 06:32:58,250 WARN conf.HiveConf: HiveConf of name hive.stats.retries.wait
 does not exist
2022-11-13 06:33:04,082 WARN metastore.ObjectStore: Version information not found in
 metastore. hive.metastore.schema.verification is not enabled so recording the schem
a version 2.3.0
2022-11-13 06:33:04,082 WARN metastore.ObjectStore: setMetaStoreSchemaVersion called
 but recording version is disabled: version = 2.3.0, comment = Set by MetaStore root
@192.168.128.130
Spark master: local[*], Application Id: local-1668339175210
spark-sql>
```

图 2-78　启动 spark-sql

（2）使用 SQL 命令。

在 spark-sql 中输入以下命令，查看数据库情况，结果如图 2-79 所示。

spark-sql> show databases;

```
spark-sql> show databases;
default
Time taken: 3.843 seconds, Fetched 1 row(s)
spark-sql>
```

图 2-79　查看数据库情况

在 spark-sql 中输入以下命令，创建 qzct 数据库，结果如图 2-80 所示。

spark-sql> create database qzct;

```
spark-sql> create database qzct;
2022-11-13 06:43:53,866 WARN metastore.ObjectStore: Failed to get d
atabase qzct, returning NoSuchObjectException
Time taken: 0.1 seconds
spark-sql> show databases;
default
qzct
Time taken: 0.085 seconds, Fetched 2 row(s)
spark-sql>
```

图 2-80　创建 qzct 数据库

在 spark-sql 中输入以下命令，创建名为 bigdata 的表，包含 classid、classname 字段，结果如图 2-81 所示。

spark-sql> use qzct;
spark-sql> create table bigdata(
　　> classid int,classname string)
　　> row format delimited fields terminated by ',';

```
spark-sql> use qzct;
Time taken: 0.051 seconds
spark-sql> create table bigdata(
        > classid int,classname string)
        > row format delimited fields terminated by ',';
2022-11-13 06:49:40,023 WARN metastore.HiveMetaStore: Location: file:/root/s
park-warehouse/qzct.db/bigdata specified for non-external table:bigdata
Time taken: 0.269 seconds
spark-sql> show tables;
qzct    bigdata false
Time taken: 0.15 seconds, Fetched 1 row(s)
spark-sql>
```

图 2-81　创建名为 bigdata 的表

3. 使用 spark-shell

由于 Spark SQL 已经集成在 spark-shell 中，因此只要启动 spark-shell 即可使用 Spark SQL 的 shell 交互接口。

spark-shell 支持 Scala 和 Python 语言。在默认情况下，Spark 自带 Scala，本书将使用 Scala 语言在 spark-shell 中进行后续的数据分析。

（1）启动 spark-shell。

在 master 节点中输入以下命令，启动 spark-shell。

[root@master ~]# spark-shell

在启动 spark-shell 的过程中，默认会创建 sc 和 spark 两个实例对象，如图 2-82 所示。在启动完成之后即可在"scala>"提示栏中输入代码，执行交互式开发。

```
[root@master ~]# spark-shell
2022-10-19 11:07:06,717 WARN util.NativeCodeLoader: Unable to load native-hadoop library for your platform... using builtin-java classes where applicable
Setting default log level to "WARN".
To adjust logging level use sc.setLogLevel(newLevel). For SparkR, use setLogLevel(newLevel).
2022-10-19 11:07:36,109 WARN util.Utils: Service 'SparkUI' could not bind on port 4040. Attempting port 4041.
Spark context Web UI available at http://master:4041
Spark context available as 'sc' (master = local[*], app id = local-1666192056939).
Spark session available as 'spark'.
Welcome to
      ____              __
     / __/__  ___ _____/ /__
    _\ \/ _ \/ _ `/ __/  '_/
   /___/ .__/\_,_/_/ /_/\_\   version 3.1.2
      /_/

Using Scala version 2.12.10 (Java HotSpot(TM) 64-Bit Server VM, Java 1.8.0_281)
Type in expressions to have them evaluated.
Type :help for more information.

scala>
```

spark-shell 中已经默认创建了两个实例对象：sc 和 spark

在这里输入代码，执行交互式开发

图 2-82　启动 spark-shell

（2）spark-shell 的常用命令。

在 spark-shell 中输入":help"命令，查看 spark-shell 的常用命令，如图 2-83 所示。

scala> :help

```
scala> :help
All commands can be abbreviated, e.g., :he instead of :help.
:completions <string>    output completions for the given string
:edit <id>|<line>        edit history
:help [command]          print this summary or command-specific help
:history [num]           show the history (optional num is commands to show)
:h? <string>             search the history
:imports [name name ...] show import history, identifying sources of names
:implicits [-v]          show the implicits in scope
:javap <path|class>      disassemble a file or class name
:line <id>|<line>        place line(s) at the end of history
:load <path>             interpret lines in a file
:paste [-raw] [path]     enter paste mode or paste a file
:power                   enable power user mode
:quit                    exit the interpreter
:replay [options]        reset the repl and replay all previous commands
:require <path>          add a jar to the classpath
:reset [options]         reset the repl to its initial state, forgetting all session entries
:save <path>             save replayable session to a file
:sh <command line>       run a shell command (result is implicitly => List[String])
:settings <options>      update compiler options, if possible; see reset
:silent                  disable/enable automatic printing of results
:type [-v] <expr>        display the type of an expression without evaluating it
:kind [-v] <type>        display the kind of a type. see also :help kind
:warnings                show the suppressed warnings from the most recent line which had any
```

图 2-83　查看 spark-shell 的常用命令

在 spark-shell 中输入":paste"命令，进入 paste 模式，从而实现大量多行代码块粘贴运行。

```
scala> :paste
```

将以下计算过程复制到 spark-shell 中，同时按"Ctrl+D"组合键，执行以下代码。

```
val w = 5
val l = 4
val area = w * l
println("面积为："+ area)
```

上述代码块实现了简单的矩形面积的计算，利用 paste 模式可以快速运行多行代码，提高操作效率，如图 2-84 所示。

```
scala> :paste
// Entering paste mode (ctrl-D to finish)

val w = 5
val l = 4
val area = w * l
println("面积为：" + area)

// Exiting paste mode, now interpreting.

面积为：20
w: Int = 5
l: Int = 4
area: Int = 20
```

图 2-84　paste 模式运行代码

任务 2.3　安装 Scala

Scala 安装

情境导入

Scala 是 Spark 常用的编程语言之一，本书后续编程使用的语言也是 Scala。在"2.2.3 Spark 集群测试"的"spark-shell 的常用命令"中，已经简单接触了 Scala 语言。Scala 可以运行在许多环境中，本任务将在 Linux 中安装 Scala。

学习目标和要求

知识与技能目标

1. 能够下载、安装配置 Scala。
2. 能够启动 Scala 进行编程测试。

素质目标

养成有条理、细致认真的工作习惯。

2.3.1 下载 Scala 安装包

本书使用的 Scala 版本是 scala-2.13.6.tgz，可以在 Scala 官方网站中查找所有版本，如图 2-85 所示。

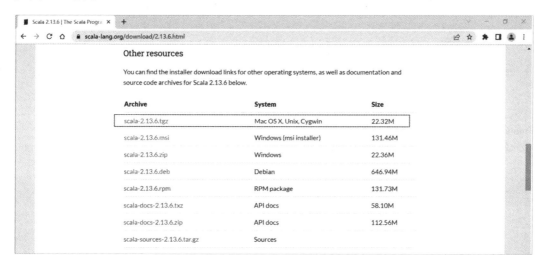

图 2-85　下载 Scala 安装包

2.3.2 Scala 安装配置

1. 上传 Scala 安装包至 master 节点的"root"目录下，解压缩并修改名称

在 master 节点中输入以下命令，操作过程与结果如图 2-86 所示。

[root@master ~]# tar -zxf scala-2.13.6.tgz -C /usr/local/
[root@master ~]# mv /usr/local/scala-2.13.6/ /usr/local/scala
[root@master ~]# ls /usr/local/

```
[root@master ~]# tar -zxf scala-2.13.6.tgz -C /usr/local/
[root@master ~]# mv /usr/local/scala-2.13.6/ /usr/local/scala
[root@master ~]# ls /usr/local/
bin    games    include         lib     libexec   scala   spark
etc    hadoop   jdk1.8.0_281    lib64   sbin      share   src
[root@master ~]#
```

图 2-86　解压缩 Scala 安装包并重命名

2. 配置环境变量，添加 Scala 相关配置

在 master 节点中输入以下命令，修改环境变量文件，添加图 2-87 所示的配置。

[root@master ~]# vi .bash_profile

```
# User specific environment and startup programs

PATH=$PATH:$HOME/bin
export JAVA_HOME=/usr/local/jdk1.8.0_281
export HADOOP_HOME=/usr/local/hadoop
export SPARK_HOME=/usr/local/spark
export SCALA_HOME=/usr/local/scala
PATH=$PATH:$JAVA_HOME/bin:$HADOOP_HOME/bin:$HADOOP_HOME/sbin:$SPARK_HOME/bin:$SPARK_HOME/sbin:$SCALA_HOME/bin
export PATH
```

图 2-87　修改环境变量文件

3. 使环境变量生效，查看 Scala 版本

在 master 节点中输入以下命令，结果如图 2-88 所示，表示安装成功。

[root@master ~]# source .bash_profile
[root@master ~]# scala -version

```
[root@master ~]# source .bash_profile
[root@master ~]# scala -version
Scala code runner version 2.13.6 -- Copyright 2002-2021, LAMP/EPFL and Lightbend, Inc.
```

图 2-88　使环境变量生效，查看 Scala 版本

4. 启动 Scala

在 master 节点中输入以下命令，启动 Scala。

[root@master ~]# scala

```
[root@master ~]# scala
Welcome to Scala 2.13.6 (Java HotSpot(TM) 64-Bit Server VM, Java 1.8.0_281).
Type in expressions for evaluation. Or try :help.

scala>
```

图 2-89　启动 Scala

由图 2-89 可知 Scala 启动成功，从下一章开始 Scala 编程练习。

脑图小结

本章通过详细的操作步骤与结果分析，介绍了 Hadoop 集群环境搭建、Spark 集群部署与使用、Scala 安装；详细阐述了各个集群环境搭建中的相关配置文件。通过脑图小结，助力学习者掌握并巩固相关知识。

第 2 章 实践环境准备

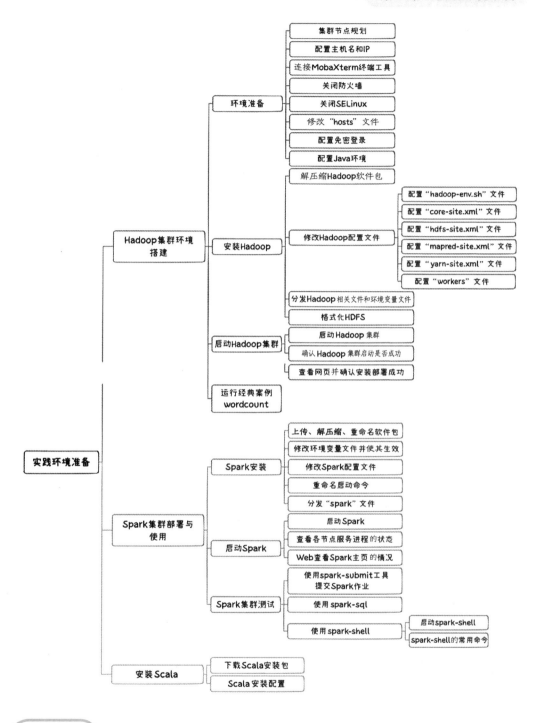

章节练习

1. 选择题

（1）能够同时启动 HDFS 和 YARN 的命令是（　　）。

A．start-dfs.sh　　　B．start-yarn.sh

C．start-all.sh　　　D．start-hadoop.sh

（2）执行 mapreduce jar 包的命令是（　　）。

A．hadoop jar　　B．java jar　　C．java -jar　　D．hdfs dfs jar

（3）负责 HDFS 数据存储的程序是（　　）。

A．NameNode　　　　　　　B．Jobtracker

C．DataNode　　　　　　　D．secondaryNameNode

（4）HDFS 中的 block 默认保存（　　）份。

A．3　　　　B．2　　　　C．1　　　　D．不确定

（5）以下不属于 Hadoop 内核组成部分的是（　　）。

A．HDFS　　　B．MapReduce

C．Hbase　　　D．YARN

2．填空题

（1）Hadoop 的部署方式有_____、_____、_____。

（2）修改_____文件，可以建立主机和 IP 地址之间的映射关系。

（3）Spark 的核心组件有_____。

（4）对于数据的批处理，通常采用编写程序、打 .jar 包的方式提交给集群执行，这需要使用 Spark 自带的_____工具。

（5）启动 spark-shell 的命令是_____。

3．判断题，在括号中填写"√"或"×"

（1）Hadoop 指定 ResourceManager 的位置在"yarn-site.xml"文件中。（　　）

（2）MapReduce 分为 map 阶段和 reduce 阶段。（　　）

（3）因为 Hadoop 是基于 Java 开发的，所以 MapReduce 只支持 Java 语言。（　　）

（4）因为 HDFS 有多个副本，所以 NameNode 是不存在单点问题的。（　　）

4．简答题

（1）简述本书安装的 3 个 Hadoop 节点的角色分布。

（2）在 Hadoop 完全分布式部署环境准备中，关闭防火墙和 SELinux 的作用是什么？

（3）简述 Hadoop 部署中各配置文件的作用。

（4）简述 Spark 部署中各配置文件的作用。

（5）简述 spark-submit 提交任务中各参数的作用。

第 3 章

学生信息处理分析

在信息化、数字化快速发展的时代，各行各业都会利用各种信息化手段对数据信息进行存储、分析与使用。在学校中，教务部门会对学生信息进行记录，为每一位学生录入唯一的学号，并分配到各个班级。有效地存储与使用学生信息，可以帮助辅导员和老师们更方便地了解并管理学生。例如，利用学号查找学生班级，统计班级基本情况，获取学生个人体质信息、特长、成绩情况等。本章将详细讲解 Scala 的基础知识，使学习者掌握用 Scala 语言对学生信息进行处理分析的方法。

Scala 是 Scalable Language 的简写，是 Spark 常用的编程语言之一，本书后续的编程均使用 Scala 语言。

Scala 语言的第一个特点是面向对象。对象是指把数据及对数据的操作方法放在一起，作为一个相互依存的整体。Scala 语言是一种形式纯净的面向对象语言，所有的值都是对象，所有的运算都是方法的调用。对象的数据类型及行为由类和特征来描述。类抽象机制的扩展通过两种途径实现：一种是子类继承，另一种是混入机制，这两种途径都能够避免多重继承的问题。

Scala 语言的第二个特点是函数式编程。Scala 语言既面向对象，又面向过程。在 Scala 语言中，函数与类、对象的地位是一样的。因此，Scala 语言的面向过程重点就是针对函数的编程，所以被称为函数式编程。函数式语言有两个主要理念：一是函数是第一级别数值，即函数和整数、字符串等的级别相同，可以作为其他函数的参数、返回值、变量存储等，这给组合函数带来了很大的便利；二是函数中的运算是将输入数值映射到输出数值，而不是对数据进行原地改动。Scala 语言提供了轻量级的语法用于定义匿名函数，支持高阶函数，允许嵌套多层函数，并支持柯里化。

Scala 语言的第三个特点是可扩展。Scala 语言的名称来自"可扩展的语言"。之所以这样命名，是因为它被设计成可以随着用户的需求来扩展的语言。它提供了隐式类和字符串插值的独特语言机制来实现可扩展。隐式类允许给已有的类型添加扩展方法，字符串插值可以让用户使用自定义的插值器进行扩展，能够很容易地以库的方式无缝添加新的语言结构。

Scala 语言的第四个特点是静态类型。Scala 语言是静态类型的语言，严格遵守变量的数据类型，拥有一个强大的类型系统，以安全、一致的方式使用抽象。它的类型系统支持泛型类、协变和逆变、标注、类型参数的上下限约束、复合类型、视图、多态方法，可以把类别和抽象类型作为对象成员，以及在引用自己时显式指定类型。

Scala 语言的第五个特点是基于 JVM 平台，可以交互操作，具有很好的兼容性和并发性。它可以复用很多 Java 的资源和库，以及跨平台运行。但是 Scala 语言不是简单地复用 Java 的元素，而是对 Java 的元素进行封装以实现自己的特性。Scala 语言可以与 Java 交互操作，利用 Scalac 编译器将源文件编译成 Java 的 class 文件运行，可以实现从中调用所有的 Java 类库，或从 Java 应用程序中调用 Scala 的代码。

接下来讲解 Scala 语言的基本概念与操作，并完成学生信息处理分析任务。

任务 3.1　班级基本情况分析

情境导入

现有一份关于某校信息工程学院所有学生基本情况的原始数据，如图 3-1 所示。数据包含 8 个字段，分别为学号、姓名、性别、年级、班级、语文成绩、数学成绩、英语成绩。辅导员想通过一位学生的学号获知其所属的班级，同时对各班级基本情况进行了解，以便后续开展工作。大数据分析老师希望李雷同学使用 Scala 语言解决这些问题。

```
21103,严林石,男,21级,大数据1班,89,97,91
21203,郑翔天,男,21级,大数据2班,87,94,91
21102,王西茜,女,21级,大数据1班,85,89,83
21201,吴倩瑶,女,21级,大数据2班,88,89,67
21104,黄雨佳,女,21级,大数据1班,90,79,84
21206,周洛晨,男,21级,大数据2班,76,77,78
21105,陈涵,女,21级,大数据1班,79,80,82
21106,沈高翔,男,21级,大数据1班,89,82,83
21110,叶嘉乐,男,21级,大数据1班,65,85,79
21108,庄伟康,男,21级,大数据1班,67,94,87
21208,陈心雨,女,21级,大数据2班,84,92,91
21111,叶聪,男,21级,大数据1班,78,88,77
21109,穆子洋,男,21级,大数据1班,91,76,78
21204,彭赟赟,女,21级,大数据2班,85,79,93
21202,蒲加敏,男,21级,大数据2班,93,93,90
21107,邹和俊,男,21级,大数据1班,77,83,88
21205,赵影,女,21级,大数据2班,80,69,95
21101,周志豪,男,21级,大数据1班,86,90,80
21207,杨溢,男,21级,大数据2班,60,90,90
```

图 3-1　原始数据

学习目标和要求

知识与技能目标

1. 掌握 Scala 数据类型、常量与变量、运算符。

2. 掌握 Scala 判断与循环、函数式编程。
3. 掌握 Scala 集合操作。

素质目标

1. 具有按照规范编写程序的意识及职业精神。
2. 养成在程序编写过程中细致入微的工作态度。

3.1.1 学生所属班级和男女生数量

一、理论基础

1. 数据类型

Scala 的数据类型与 Java 的相同，但是 Scala 中的数据类型都是对象，即 Scala 没有 Java 中的原生类型；因此，Scala 可以对数字等基础类型调用方法。Scala 的数据类型包括 Byte、Short、Int、Long、Float、Double、Char、String、Boolean、Unit、Null、Nothing、Any、AnyRef，具体描述如表 3-1 所示。

表 3-1 Scala 数据类型及描述

数据类型	类型描述
Byte	表示 8 位有符号补码整数，其数值区间为 -128～127
Short	表示 16 位有符号补码整数，其数值区间为 -32768～32767
Int	表示 32 位有符号补码整数，其数值区间为 -2147483648～2147483647，只能保存整数
Long	表示 64 位有符号补码整数，其数值区间为 -9223372036854775808～9223372036854775807
Float	表示 32 位 IEEE754 标准的单精度浮点数，如果浮点数后面有 f 或者 F 后缀，则表示是 Float 类型的，否则是 Double 类型的
Double	表示 64 位 IEEE754 标准的双精度浮点数
Char	表示 16 位无符号 Unicode 字符，区间值为 U+0000～U+FFFF
String	表示字符序列，即字符串，在使用时需要使用双引号包含一系列字符
Boolean	其值为 true 或 false
Unit	表示无值，和其他语言中的 void 等同，用于不返回任何结果的方法的数据类型。Unit 只有一个实例值，写成 ()
Null	null 或空引用
Nothing	Nothing 类型在 Scala 类层级的最底端，是任何其他类型的子类型
Any	Any 是所有其他类型的超类
AnyRef	AnyRef 是所有引用类（Reference Class）的基类

由于 Scala 支持数据类型推断，因此在定义变量时多数可以不指明数据类型，而由 Scala 运行环境自动给出变量的数据类型。在 scala 命令行中输入以下命令，使用类型推断获得输出数据的数据类型，如图 3-2 所示。

```
scala> 2.5+3
val res0: Double = 5.5
```

图 3-2　使用类型推断获得输出数据的数据类型

由图 3-2 可知，在 Double 和 Int 类型的数据相加之后，Scala 会自动确定输出结果的数据类型是 Double 类型。

2. 常量与变量

在 Scala 中，数据可以分为常量和变量两种类型。常量不可变，意味着其值一旦声明了就不能重新赋值更改。变量是在程序运行过程中其值可能改变的量。

（1）定义方法。

在 Scala 中，定义常量的关键字是 val，语法格式如下：

val 常量名称 : 数据类型 = 初始值

关键字 val 后面紧接着常量名称、数据类型、等号和初始值。常量的命名遵守 Scala 标识符规则：以字母或者下画线开头，后接字母、数字、下画线；以操作符开头，且只包含操作符；用反引号包括的任意字符串。

在 Scala 程序中，通常建议使用常量关键字 val 来定义常量。因为在类似 Spark 集群的大型复杂系统中，需要大量的网络传输数据，如果使用变量 var，则可能会被错误地更改。

val 一经初始化就无法再进行修改，否则会报错。在 scala 命令行中输入以下命令，定义常量 a，其数据类型是 Int，初始值为 1。重新将 a 赋值为 3，则会报错，如图 3-3 所示。

val a:Int=1

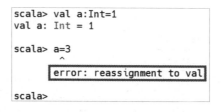

图 3-3　使用关键字 val 定义常量

在 Scala 中，定义变量的关键字是 var，语法格式如下：

var 变量名称 : 数据类型 = 初始值

与常量定义一样，在关键字 var 后紧接着变量名称、数据类型、等号和初始值。变量的命名规则同样要遵守 Scala 标识符规则。与常量不同的是，变量被定义后可以重新赋值。在 scala 命令行中输入以下命令，定义变量 b，其数据类型是 Int，初始值为 2。重新将 b 赋值为 3，赋值成功。但是将 b 重新赋值为 school 就会报错，因为只能将与变量同类型的值赋给变量，而此处 school 为字符串，如图 3-4 所示。

var b:Int=2

```
scala> var b:Int=2
var b: Int = 2

scala> b = 3
// mutated b

scala> b
val res0: Int = 3

scala> b="school"
       ^
       error: type mismatch;
        found   : String("school")
        required: Int
```

图 3-4　使用关键字 var 定义变量

（2）数据类型推断。

Scala 具备数据类型推断的功能，因此在定义常量或者变量时，可以不用特地说明数据的类型，即可以不在常量或变量名称后面添加":type"数据类型项。定义常量 c 和 d，分别赋值为 3 和 school；定义变量 e 和 f，分别赋值为 4 和 grade，如图 3-5 所示。在定义常量和变量，且没有指定数据类型的情况下，Scala 会根据初始值进行数据类型的推断。

```
scala> val c=3
val c: Int = 3

scala> val d="school"
val d: String = school

scala> var e=4
var e: Int = 4

scala> var f="grade"
var f: String = grade
```

图 3-5　数据类型推断

（3）多变量声明。

Scala 支持多个常量或变量的声明，以变量声明为例，声明方法如下。

若声明的多个变量初始值相同，则只需用逗号分隔，命令如下，结果如图 3-6 所示。

var a, b = 2

```
scala> var a, b = 2
var a: Int = 2
var b: Int = 2
```

图 3-6　多变量声明方法（1）

若声明的多个变量初始值不相同，则需将它们声明为元组（元组将在后续集合中详细讲解），命令如下，结果如图 3-7 所示。

var (a, b, c) = (1, 2, 3)

```
scala> var (a, b, c) = (1, 2, 3)
var a: Int = 1
var b: Int = 2
var c: Int = 3
```

图 3-7　多变量声明方法（2）

3. 运算符

Scala 中的运算符即方法，方法即运算符，它其实是普通方法调用的另一种表现形式。运算符的使用其实就是隐含地调用对应的方法，也可以通过"值.运算符"的方式实现。Scala 运算符包含算数、关系、逻辑、赋值和位运算符。表 3-2 所示为 Scala 支持的各种运算符。

表 3-2 Scala 运算符

运算符类别	运算符	描述	举例
算数运算符举例：假设 a=1，b=2			
算数运算符	+	加号	a+b 的运算结果为 3
	-	减号	a-b 的运算结果为 -1
	*	乘号	a*b 的运算结果为 2
	/	除号	b/a 的运算结果为 2
	%	取余	b%a 的运算结果为 0
关系运算符举例：假设 a=1，b=2			
关系运算符	==	等于	a==b 的运算结果为 false
	>	大于	a>b 的运算结果为 false
	<	小于	a<b 的运算结果为 true
	>=	大于或等于	a>=b 的运算结果为 false
	<=	小于或等于	a<=b 的运算结果为 true
	!=	不等于	a!=b 的运算结果为 true
逻辑运算符举例：假设 a=1，b=0			
逻辑运算符	&&	与，两个条件均成立则为真，否则为假	a&&b 的运算结果为 false
	||	或，两个条件中有一个成立则为真，否则为假	a||b 的运算结果为 true
	!	非，对当前条件取反	!(a&&b) 的运算结果为 true
位运算符举例：假设 a=0011 1100，b=0000 1101			
位运算符	&	按位与，数据按照二进制位进行运算，两位均为 1 则结果为 1，否则为 0	a&b 的运算结果为 0000 1100
	|	按位或，数据按照二进制位进行运算，两位中有一个为 1 则结果为 1	a|b 的运算结果为 0011 1101
	^	按位异或，数据按照二进制位进行运算，两位不同时结果为 1，相同时为 0	a^b 的运算结果为 0011 0001
	~	按位取反，数据按照二进制位进行运算，是 1 则变为 0，是 0 则变成 1	~a 的运算结果为 1100 0011
	<<	数据按照二进制位进行运算，左移	a << 2 的运算结果为 1111 0000
	>>	数据按照二进制位进行运算，右移	a >> 2 的运算结果为 0000 1111
	>>>	数据按照二进制位进行运算，无符号右移	a >>> 2 的运算结果为 0000 1111

续表

运算符类别	运算符	描述	举例
赋值运算符	=	指定右边操作结果赋值给左边	c=a+b，将 a+b 的运算结果赋值给 c
	+=	左右两边相加后赋值给左边	c+=a 即 c=c+a
	-=	左右两边相减后赋值给左边	c-=a 即 c=c-a
	*=	左右两边相乘后赋值给左边	c*=a 即 c=c*a
	/=	左右两边相除后赋值给左边	c/=a 即 c=c/a
	%=	左右两边求余后赋值给左边	c%=a 即 c=c%a
	<<=	按位左移后赋值给左边	c<<=a 即 c=c<<a
	>>=	按位右移后赋值给左边	c>>=a 即 c=c>>a
	&=	按位与运算后赋值给左边	c&=a 即 c=c&a
	|=	按位或运算后赋值给左边	c|=a 即 c=c|a
	^=	按位异或运算后赋值给左边	c^=a 即 c=c^a

4. 数组

Scala 提供了一种数据结构叫作数组，数组是用于存储相同类型变量的集合。数组分为定长数组（Array）和变长数组（ArrayBuffer）。定长数组就是长度不变的数组，使用 Array 进行声明。不定长数组就是长度可变的数组，使用 ArrayBuffer 进行声明。需要注意的是，Scala 默认创建的是定长数组，当创建变长数组时需要先导入相关依赖包。

（1）声明与定义方式。

在 Scala 中，声明与定义一个数组有 3 种方式，即使用 new 的完整形式、使用 new 的简写形式，以及定义并初始化的方式。

使用 new 的完整形式，其语法如下：

var arrname:Array[type] = new Array[type](size)

定义一个数组 arrname，该数组拥有 size 个元素，每个元素的类型都是 type。在 scala 命令行中输入以下命令，进行数组的定义，结果如图 3-8 所示。

var arr:Array[Int]=new Array[Int](4)
arr(0)=1;arr(1)=3;arr(2)=12;arr(3)=10
arr

```
scala> var arr:Array[Int]=new Array[Int](4)
var arr: Array[Int] = Array(0, 0, 0, 0)

scala> arr(0)=1;arr(1)=3;arr(2)=12;arr(3)=10

scala> arr
val res11: Array[Int] = Array(1, 3, 12, 10)
```

图 3-8 使用 new 的完整形式定义数组

使用 new 的简写形式，其语法如下：

var arrname = new Array[type](size)

在 scala 命令行中输入以下命令，进行数组的定义，结果如图 3-9 所示。

```
var arr=new Array[Int](4)
arr(0)=1;arr(1)=3;arr(2)=12;arr(3)=10
arr
```

使用定义并初始化的方式，其语法如下：

```
var arrname = Array(element1, element2, element3)
```

定义一个数组 arrname，该数组中的元素为 element1、element2 和 element3。在 scala 命令行中输入以下命令，进行数组的定义，结果如图 3-10 所示。

```
var arr = Array(1,3,12,10)
```

```
scala> var arr=new Array[Int](4)
var arr: Array[Int] = Array(0, 0, 0, 0)

scala> arr(0)=1;arr(1)=3;arr(2)=12;arr(3)=10

scala> arr
val res14: Array[Int] = Array(1, 3, 12, 10)
```

```
scala> var arr=Array(1,3,12,10)
var arr: Array[Int] = Array(1, 3, 12, 10)
```

图 3-9　使用 new 的简写形式定义数组　　　　图 3-10　定义并初始化数组

前面讲解的是定长数组的定义与声明方式，变长数组的定义与声明方式与定长数组的相同，但是需要先导入 "import scala.collection.mutable.ArrayBuffer" 依赖包。

在 scala 命令行中输入以下命令，进行变长数组的定义与声明操作，结果如图 3-11 所示。

```
import scala.collection.mutable.ArrayBuffer
var arrbuffer = ArrayBuffer(1, 5, 3, 7, 10, 2)
// 增加一个元素
arrbuffer += 20
// 增加一个数组集合
arrbuffer ++= Array(50, 60)
// 删除最后 3 个元素
arrbuffer.dropRightInPlace(3)
// 清空数组
arrbuffer.clear()
```

```
scala> import scala.collection.mutable.ArrayBuffer
import scala.collection.mutable.ArrayBuffer

scala> var arrbuffer = ArrayBuffer(1, 5, 3, 7, 10, 2)
var arrbuffer: scala.collection.mutable.ArrayBuffer[Int] = ArrayBuffer(1, 5, 3, 7, 10, 2)

scala> arrbuffer += 20
val res21: scala.collection.mutable.ArrayBuffer[Int] = ArrayBuffer(1, 5, 3, 7, 10, 2, 20)

scala> arrbuffer ++= Array(50, 60)
val res22: scala.collection.mutable.ArrayBuffer[Int] = ArrayBuffer(1, 5, 3, 7, 10, 2, 20, 50, 60)

scala> arrbuffer.dropRightInPlace(3)
val res23: scala.collection.mutable.ArrayBuffer[Int] = ArrayBuffer(1, 5, 3, 7, 10, 2)

scala> arrbuffer.clear()

scala> arrbuffer
val res25: scala.collection.mutable.ArrayBuffer[Int] = ArrayBuffer()
```

图 3-11　变长数组的定义与声明操作

（2）基本操作方法。

在 Scala 中，数组是一种重要的数据结构，拥有一些基本的操作方法，如表 3-3 所示。

表 3-3 数组的基本操作方法

基本操作	描　　述
arr.sum	对于 Int 数据类型的数组，返回数组各元素的和
arr.max	对于 Int 数据类型的数组，返回数组各元素的最大值
arr.min	对于 Int 数据类型的数组，返回数组各元素的最小值
arr.length	返回数组的长度
arr.sorted.toBuffer	对于 Int 数据类型的数组，返回数组各元素由小到大排序的结果
arr.reverse.toBuffer	对于 Int 数据类型的数组，返回数组各元素倒序的结果
arr.contains(x)	判断数组中是否包含 x 元素
arr.isEmpty	判断数组是否为空

在 scala 命令行中输入以下命令，对前面定义的数组 arr 进行数组的基本操作，结果如图 3-12 所示。

arr.sum
arr.max
arr.min
arr.length
arr.sorted.toBuffer
arr.reverse.toBuffer
arr.contains(5)
arr.isEmpty

```
scala> arr.sum
val res17: Int = 26

scala> arr.max
val res18: Int = 12

scala> arr.min
val res19: Int = 1

scala> arr.length
val res20: Int = 4

scala> arr.sorted.toBuffer
val res21: scala.collection.mutable.Buffer[Int] = ArrayBuffer(1, 3, 10, 12)

scala> arr.reverse.toBuffer
val res22: scala.collection.mutable.Buffer[Int] = ArrayBuffer(10, 12, 3, 1)

scala> arr.contains(5)
val res23: Boolean = false

scala> arr.isEmpty
val res24: Boolean = false
```

图 3-12 数组的基本操作

由图 3-12 可知，数组 arr 中各元素的和为 26，最大值为 12，最小值为 1，数组的长度为 4，即数组中有 4 个元素。输出结果为对数组中各元素从小到大进行排序，以及使用 reverse.toBuffer 进行倒序显示。因为数组中没有数字 5，所以 arr.contains(5) 的结果为 false。又因为数组中是有元素的，所以使用 isEmpty() 方法的结果也是 false。

5. 判断与循环

Scala 中的判断与循环语法与其他计算机语言一样。if 判断可以对数据进行过滤判断处理，while 和 for 循环可以在某一条件下将某段代码进行重复执行。

（1）if 判断。

if 判断有 4 种形式，分别为 if 语句、if...else 语句、if...else if...else 语句和 if...else 嵌套语句。

if 语句的语法格式如下：

```
if( 判断条件 )
{
 // 如果判断条件为 true 则执行该语句块
}
```

例如，判断一个变量是否等于 100，若是则打印"满分"，使用 if 语句完成。在 scala 命令行中输入以下命令，结果如图 3-13 所示。

```
var score=100
if(score==100){
    println(" 满分 ")
    }
```

```
scala> var score=100
var score: Int = 100

scala> if(score==100){
     | println("满分")
     | }
满分
```

图 3-13 if 语句举例

if...else 语句适合一个判断条件的情况，其语法格式如下：

```
if( 判断条件 ){
 // 如果判断条件为 true 则执行该语句块
}else{
 // 如果判断条件为 false 则执行该语句块
}
```

例如，判断一个变量是否大于 90，若是则打印"优秀"，否则打印"其他"，使用 if...else 语句完成。在 scala 命令行中输入以下命令，结果如图 3-14 所示。

```
var score=95
if(score>90){
    println(" 优秀 ")
    }else{
    println(" 其他 ")
    }
```

第 3 章 学生信息处理分析

```
scala> var score=95
var score: Int = 95

scala> if(score>90){
     | println("优秀")
     | }else{
     | println("其他")
     | }
优秀
```

图 3-14 if...else 语句举例

if...else if...else 语句适合多个判断条件的情况，其语法格式如下：

if(判断条件 1){
　// 如果判断条件 1 为 true 则执行该语句块
}else if(判断条件 2){
　// 如果判断条件 2 为 true 则执行该语句块
}else{
　// 如果以上判断条件都为 false 则执行该语句块
}

例如，判断一个变量是否大于 90，若是则打印"优秀"，否则判断是否大于 80；若大于 80 则打印"良好"，否则判断是否大于 60；若大于 60 则打印"及格"，否则打印"不及格"，使用 if...else if...else 语句完成。在 scala 命令行中输入以下命令，结果如图 3-15 所示。

```
var score=85
if(score>90){
    print(" 优秀 ")
    }else if(score>80){
    print(" 良好 ")
    }else if(score>60){
    print(" 及格 ")
    }else{
    print(" 不及格 ")
    }
```

```
scala> var score=85
var score: Int = 85

scala> if(score>90){
     | print("优秀")
     | }else if(score>80){
     | print("良好")
     | }else if(score>60){
     | print("及格")
     | }else{
     | print("不及格")
     | }
良好
```

图 3-15 if...else if...else 语句举例

if...else 嵌套语句用于实现在 if 语句内嵌入一个或多个 if 语句，其语法格式如下：

if(判断条件 1){
　// 如果判断条件 1 为 true 则执行该语句块

```
if( 判断条件 2){
  // 如果判断条件 2 为 true 则执行该语句块
  }
}
```

例如，判断一个变量是否大于 70 且小于 80，若是则打印"中等"，使用 if...else 嵌套语句实现，在 scala 命令行中输入以下命令，结果如图 3-16 所示。

```
var score=75
if(score>70){
    if(score<80){
    print(" 中等 ")
    }
    }
```

```
scala> var score=75
var score: Int = 75

scala> if(score>70){
     | if(score<80){
     | print("中等")
     | }
     | }
中等
```

图 3-16　if...else 嵌套语句举例

（2）循环。

Scala 中的循环有 3 种方式，分别为 while 循环、do...while 循环和 for 循环。

使用 while 循环，当给定的条件表达式为 true 时，则重复执行循环体内的代码块，否则退出循环，继续执行紧接着循环的下一条语句，其语法格式如下：

```
while( 条件表达式 )
{
  代码块
}
```

例如，假设变量 a 的初始值为 4，如果 a 大于 0 则将 a 减 1 并打印，使用 while 循环完成。在 scala 命令行中输入以下命令，结果如图 3-17 所示。

```
var a = 4
while (a > 0) {
a -= 1
printf("a is %d\n",a)
}
```

```
scala> var a = 4
var a: Int = 4

scala> while (a > 0) {
     | a -= 1
     | printf("a is %d\n",a)
     | }
a is 3
a is 2
a is 1
a is 0
```

图 3-17　while 循环举例

使用 do...while 循环，在循环的尾部检查是否符合条件表达式，此循环会确保至少执行一次循环。其语法格式如下：

do {
　　代码块
} while(条件表达式)

例如，假设变量 a 的初始值为 0，对变量 a 进行加 1 循环，直到 a 大于或等于 5 为止，使用 do...while 循环完成。在 scala 命令行中输入以下命令，结果如图 3-18 所示。

```
var a = 0
do {
    a += 1
    println(a)
}while (a<5)
```

```
scala> var a = 0
var a: Int = 0

scala> do {
    |     a += 1
    |     println(a)
    | }while (a<5)
1
2
3
4
5
```

图 3-18　do...while 循环举例

for 循环的语法格式如下，其中 range 是一个有起点和终点的值，可以使用 to 或 until 关键字来传递范围。

for(var x <- range){
　　代码块
}

例如，使用 to 关键字进行从 1 到 10 的 for 循环输出，在 scala 命令行中输入以下命令，结果如图 3-19 所示。

```
for( a <- 1 to 10 ){
    println(a)
  }
```

```
scala> for( a <- 1 to 10 ){
    |         println(a)
    |       }
1
2
3
4
5
6
7
8
9
10
```

图 3-19　使用 to 关键字的 for 循环举例

例如，使用 until 关键字进行从 1 到 10 的 for 循环，在 scala 命令行中输入以下命令，结果如图 3-20 所示。

```
for( a <- 1 until 10 ){
   println(a)
  }
```

```
scala> for( a <- 1 until 10 ){
     | println(a)
     | }
1
2
3
4
5
6
7
8
9
```

图 3-20　使用 until 关键字的 for 循环举例

由两种关键字的 for 循环结果可知，当使用 to 关键字时，区间范围包括起点和终点；而当使用 until 关键字时，区间范围不包括终点。

例如，for 循环可以通过传递条件表达式来过滤数据，在 scala 命令行中输入以下命令，过滤出 1 到 10 之间能被 3 整除的数字，结果如图 3-21 所示。

```
for(a<-1 to 10;if a%3==0){
   println(a)
  }
```

```
scala> for(a<-1 to 10;if a%3==0){
     | println(a)
     | }
3
6
9
```

图 3-21　for 循环过滤数据举例

例如，可以在 for 循环中使用分号分隔多个范围，在这种情况下，循环将遍历给定范围中的所有逻辑。在 scala 命令行中输入以下命令，实现 for 循环嵌套，结果如图 3-22 所示。

```
for(i <- 1 to 2;j <- 1 to 2){
   println("(" + i + "," + j + ")")
  }
```

```
scala> for(i <- 1 to 2;j <- 1 to 2){
     | println("(" + i + "," + j + ")")
     | }
(1,1)
(1,2)
(2,1)
(2,2)
```

图 3-22　for 循环嵌套举例

6. 函数式编程

（1）函数的定义和调用。

Scala 支持函数式编程，函数是 Scala 语言的核心。函数的定义方法如下：

Scala 函数式编程

```
def 函数名 [ 参数列表 ]:[ 返回值类型 ]={
  函数体
  return[ 表达式 ]
}
```

定义函数的关键字是 def，关键字后面是函数名。参数列表和返回值类型是可选项。参数列表中包含每个参数的名称和参数类型，以冒号分隔。返回值类型可以是任何数据类型，若函数无返回值，则返回值类型为 Unit。等号后面是函数体，最后以 return 关键字指明返回值，此关键字可以不写。

定义两个数相减的函数，在 scala 命令行中输入以下命令，结果如图 3-23 所示。

```
def minus(a:Int,b:Int):Int={a-b}
```

```
scala> def minus(a:Int,b:Int):Int={a-b}
def minus(a: Int, b: Int): Int
```

图 3-23　定义函数

在函数定义完成之后就可以进行调用了，函数调用的方法如下：

函数名 [参数列表]

调用前述定义的 minus() 函数，在 scala 命令行中输入以下命令，结果如图 3-24 所示。

```
minus(2,1)
```

```
scala> minus(2,1)
val res6: Int = 1
```

图 3-24　调用函数

（2）匿名函数。

匿名函数是指不含函数名称的函数。使用"=>"定义，"=>"的左边为参数列表，"=>"的右边为函数体表达式。

在 scala 命令行中输入以下命令，进行匿名函数的定义。

```
// 匿名函数 1
(x:Int)=>x*x
// 匿名函数 2
(x:Int)=>{println(x);x*x}
// 匿名函数赋值
val fun=(x:Int)=>{println(x);x*x}
// 匿名函数调用
fun(2)
```

```
val f1: Int => Int = $Lambda$1130/1786758337@270097ce
val res7: Int = 4

scala> (x:Int)=>x*x
val res8: Int => Int = $Lambda$1131/1311057965@3b29d36c

scala> (x:Int)=>{println(x);x*x}
val res9: Int => Int = $Lambda$1132/292486757@118879d2

scala> val fun=(x:Int)=>{println(x);x*x}
val fun: Int => Int = $Lambda$1133/997361164@7f1b2f07

scala> fun(2)
2
val res10: Int = 4
```

图 3-25　匿名函数的定义

由图 3-25 可知，匿名函数虽然没有函数名，但是可以先将其赋值给一个常量或者变量，然后通过常量或者变量名进行函数调用。fun(2) 就是调用了赋值给 fun 的匿名函数。

（3）高阶函数。

高阶函数是指使用其他函数作为参数，或者使用函数作为输出结果的函数。

定义一个使用其他函数作为参数的高阶函数，在 scala 命令行中输入以下命令，结果如图 3-26 所示。

```
// 定义一个函数 fun()，实现两个整数相加
def fun(a:Int,b:Int):Int=a+b
// 定义高阶函数 funsum()。其参数列表中有两个参数，一个是匿名函数 f，此匿名函数的参数是两个
Int 类型数据，返回值也是 Int 类型数据；另一个是字符串 s。在此高阶函数的函数体中，给匿名函数 f
传递两个参数，并赋值给常量 scoresum，最后返回字符串 s+scoresum
def funsum(f:(Int,Int)=>Int,s:String):String={
  // 给匿名函数传递参数
  val scoresum=f(95,90)
  // 返回值
  s+scoresum
  }
// 调用高阶函数
println( funsum(fun," 总成绩是 "))
// 也可以自己给函数 f 重新写一个匿名函数体
println( funsum((a:Int,b:Int)=>a+b," 总成绩是 "))
```

```
scala> def fun(a:Int,b:Int):Int=a+b
def fun(a: Int, b: Int): Int

scala> def funsum(f:(Int,Int)=>Int,s:String):String={
     |      //给匿名函数传参
     |      val scoresum=f(95,90)
     |      //返回值
     |      s+scoresum
     |      }
def funsum(f: (Int, Int) => Int, s: String): String

scala> println( funsum(fun,"总成绩是"))
总成绩是185

scala> println( funsum((a:Int,b:Int)=>a+b,"总成绩是"))
总成绩是185
```

图 3-26　使用其他函数作为参数的高阶函数

定义一个使用函数作为输出结果的高阶函数，在 scala 命令行中输入以下命令。定义一个名称为 dosquare 的高阶函数，其函数体 (x:Int)=>x*x 是一个匿名函数，调用 dosquare() 计算 2 的平方值，结果如图 3-27 所示。

```
def dosquare={
   (x:Int)=>x*x
}
dosquare(2)
```

图 3-27　使用函数作为输出结果的高阶函数

（4）闭包。

闭包是一个函数，可以访问某个函数中局部变量的另一个函数。在 scala 命令行中输入以下命令，定义一个闭包，结果如图 3-28 所示。

```
val more=2
def addmore= (x: Int) => x + more
addmore(3)
```

图 3-28　闭包定义与调用

由图 3-28 可知，在匿名函数 addmore() 中，x 是一个绑定变量，在该函数的上下文中对它有明确的定义，即它被定义为该函数的唯一参数；而 more 是一个自由变量，这个变量定义在函数外部。这样定义的函数 addmore() 就成了一个闭包，因为它引用了在函数外部定义的 more 变量。定义 addmore() 函数的过程就是"捕获"其自由变量从而"闭合"该函数的过程。

（5）函数柯里化。

在函数式编程中，将接收多个参数的函数转化为接收单个参数的函数，这个转化过程就是函数柯里化。函数柯里化本身也用到了闭包。在 scala 命令行中输入以下命令，定义一个函数，对两个整数进行相乘操作，对比常规未柯里化、闭包、函数柯里化 3 种定义方式。

```
// 未柯里化方式
def mul(x: Int, y: Int) = x * y
mul(2, 2)
```

```
// 闭包方式
def mulclosure(x: Int) = (y: Int) => x * y
println(mulclosure(3)(4))
// 函数柯里化方式
def mulcurry(x: Int)(y:Int) = x * y
println(mulcurry(5)(6))
```

在函数柯里化之后，mulcurry(5)(6) 表示依次调用两个普通函数，第一次使用 5，返回一个函数类型的值；第二次使用 6，调用返回的函数类型的值进行计算，如图 3-29 所示。

```
scala> def mul(x: Int, y: Int) = x * y
def mul(x: Int, y: Int): Int

scala> mul(2, 2)
val res14: Int = 4

scala> def mulclosure(x: Int) = (y: Int) => x * y
def mulclosure(x: Int): Int => Int

scala> println(mulclosure(3)(4))
12

scala> def mulcurry(x: Int)(y:Int) = x * y
def mulcurry(x: Int)(y: Int): Int

scala> println(mulcurry(5)(6))
30
```

图 3-29　函数柯里化

（6）嵌套函数。

嵌套函数即在函数内部定义函数，其中内部函数又称为局部函数。在 scala 命令行中输入以下命令，定义一个 mul1() 函数，其参数列表中有 a、b 和 c 这 3 个整型参数。在函数 mul1() 内部定义一个局部函数 mul2()，完成两个数据的相乘操作。接着嵌套调用 mul2() 函数，完成 3 个参数的相乘操作，结果如图 3-30 所示。

```
def mul1(a:Int, b:Int, c:Int) = {
  // 定义局部函数
  def mul2(x:Int, y:Int) = {
      x*y
  }
  // 调用局部函数
  mul2(a, mul2(b, c))
}
// 调用外部函数
mul1(2, 2, 2)
```

```
scala> def mul1(a:Int, b:Int, c:Int) = {
     |     def mul2(x:Int, y:Int) = {
     |         x*y
     |     }
     |     mul2(a, mul2(b, c))
     | }
def mul1(a: Int, b: Int, c: Int): Int

scala> mul1(2, 2, 2)
val res0: Int = 8
```

图 3-30　嵌套函数

二、编程分析实现

（1）判断学生所属班级。

完成此任务首先需要用数组存储各个班级学生的学号，然后定义函数识别某个学生的学号并获取其所属班级。由于数据较多，并且还没有介绍读取文件的方法，因此此任务将截取部分原始数据进行操作。

在 scala 命令行中输入以下命令，进行班级数组的定义，结果如图 3-31 所示。

```
val bigdata1=Array(21101,21102,21103,21104,21105,21106,21107,21108)
val bigdata2=Array(21201,21202,21203,21204,21205,21206,21207,21208)
```

```
scala> val bigdata1=Array(21101,21102,21103,21104,21105,21106,21107,21108)
val bigdata1: Array[Int] = Array(21101, 21102, 21103, 21104, 21105, 21106, 21107, 21108)

scala> val bigdata2=Array(21201,21202,21203,21204,21205,21206,21207,21208)
val bigdata2: Array[Int] = Array(21201, 21202, 21203, 21204, 21205, 21206, 21207, 21208)
```

图 3-31　定义班级数组

在 scala 命令行中输入以下命令，定义 distinguish() 函数，利用 if…else if…else 语句识别学生学号并判断所属班级，结果如图 3-32 所示。

```
def distinguish(x:Int)={
   if(bigdata1.contains(x)){
   println(" 此学生是大数据 1 班的学生 ")
   }else if(bigdata2.contains(x)){
   println(" 此学生是大数据 2 班的学生 ")
   }else{
   println(" 此学生是其他分院的学生 ")
   }
}
distinguish(21203)
```

```
scala> def distinguish(x:Int)={
     |         if(bigdata1.contains(x)){
     |         println("此学生是大数据1班的学生")
     |         }else if(bigdata2.contains(x)){
     |         println("此学生是大数据2班的学生")
     |         }else{
     |         println("此学生是其他分院的学生")
     |         }
     | }
def distinguish(x: Int): Unit

scala> distinguish(21203)
此学生是大数据2班的学生
```

图 3-32　判断学生所属班级

（2）统计男女生数量。

以部分数据为例进行统计。在 scala 命令行中输入以下命令，定义函数 count(sex: String)，统计男女生人数。在 count(sex:String) 函数中利用数组存储数据信息，通过 for 循环遍历数组，利用 if 语句判断数组中元素是否包含指定性别，并进行 sum 计数，结果如图 3-33 所示。

```
def count(sex:String):Unit={
    val data=Array("21103, 严林石 , 男 ,21 级 , 大数据 1 班 ,89,97,91",
    "21203, 郑翔天 , 男 ,21 级 , 大数据 2 班 ,87,94,91",
    "21102, 王西茜 , 女 ,21 级 , 大数据 1 班 ,85,89,83")
    var sum = 0
    for(x<-data;if x.contains(sex)){
    sum+=1}
    println(sum)
    }
count(" 男 ")
```

图 3-33　统计男女生数量

3.1.2　以班级为单位整理学生信息

Scala 列表与 Set 集合

一、理论基础

1. 列表

列表（List）与数组类似，所有元素的类型都相同。它们的区别首先是列表是不可变的，其次是列表具有递归的结构而数组不具有。

（1）定义列表。

定义不可变列表的语法格式如下：

// 创建一个包含值 1, 值 2, 值 3, 值 ... 的不可变列表
var/val 变量名 :List[元素类型]=List[元素类型](值 1, 值 2, 值 3, 值 ...)
// 使用 "::" 拼接方式来创建列表，必须在最后添加一个 Nil
val/var 变量名 = 值 1 :: 值 2 :: Nil

在 scala 命令行中输入以下命令，用两种方式定义一个不可变列表，结果如图 3-34 所示。

val a = List(1,3,4,6,9,10)
val a = 1 :: 3 :: 4 :: 6 :: 9 :: 10 :: Nil

图 3-34　定义不可变列表

定义可变列表的语法格式如下：

```
// 导入相关依赖包
import scala.collection.mutable.ListBuffer
// 创建一个空的可变列表
val/var 变量名 :ListBuffer[ 类型 ]=new ListBuffer[ 类型 ]()
// 创建一个包含值 1, 值 2, 值 3... 的可变列表
val/var 变量名 = ListBuffer( 值 1, 值 2, 值 3...)
```

在 scala 命令行中输入以下命令，用两种方式定义一个可变列表，结果如图 3-35 所示。

```
import scala.collection.mutable.ListBuffer
// 创建一个空的、整型可变列表
val b = ListBuffer[Int]()
// 创建一个包含 1,3,5,7,9 的可变列表
val b = ListBuffer(1,3,5,7,9)
```

```
scala> import scala.collection.mutable.ListBuffer
import scala.collection.mutable.ListBuffer

scala> val b = ListBuffer[Int]()
val b: scala.collection.mutable.ListBuffer[Int] = ListBuffer()

scala> val b = ListBuffer(1,3,5,7,9)
val b: scala.collection.mutable.ListBuffer[Int] = ListBuffer(1, 3, 5, 7, 9)
```

图 3-35　定义可变列表

（2）列表的常用操作方法。

列表有许多常用的操作方法，表 3-4 所示为可变和不可变列表的常用操作方法。

表 3-4　列表的常用操作方法

方法（对列表 a 和 b 进行操作）	描　　述
a.isEmpty	判断列表是否为空
a++b	拼接两个列表
a.head	获取列表的首个元素
a.tail	获取列表中除首个元素以外的剩余部分
a.reverse	反转列表
a.take(num)	获取指定个数的前缀
a.drop(num)	删除从左边开始的 num 个元素
a.toString	转换字符串
a.toArray	将列表转换为数组
List.concat(a,b)	合并两个列表
a.intersect(b)	对两个列表取交集
a.diff(b)	对两个列表取差集
a(索引值)	获取列表中指定索引值的元素

在 scala 命令行中输入以下命令，对前面定义的不可变列表 a 和可变列表 b 执行表 3-4

所示的操作，结果如图 3-36 所示。

```
import scala.collection.mutable.ListBuffer
val a = List(1,3,4,6,9,10)
val b = ListBuffer(1,3,5,7,9)
a.isEmpty
a++b
a.head
a.tail
a.reverse
a.take(2)
a.drop(2)
a.toString
a.toArray
List.concat(a,b)
a.intersect(b)
a.diff(b)
a(1)
```

```
scala> import scala.collection.mutable.ListBuffer
import scala.collection.mutable.ListBuffer

scala> val a = List(1,3,4,6,9,10)
val a: List[Int] = List(1, 3, 4, 6, 9, 10)

scala> val b = ListBuffer(1,3,5,7,9)
val b: scala.collection.mutable.ListBuffer[Int] = ListBuffer(1, 3, 5, 7, 9)

scala> a.isEmpty
val res58: Boolean = false

scala> a++b
val res59: List[Int] = List(1, 3, 4, 6, 9, 10, 1, 3, 5, 7, 9)

scala> a.head
val res60: Int = 1

scala> a.tail
val res61: List[Int] = List(3, 4, 6, 9, 10)

scala> a.reverse
val res62: List[Int] = List(10, 9, 6, 4, 3, 1)

scala> a.take(2)
val res63: List[Int] = List(1, 3)

scala> a.drop(2)
val res64: List[Int] = List(4, 6, 9, 10)

scala> a.toString
val res65: String = List(1, 3, 4, 6, 9, 10)

scala> a.toArray
val res66: Array[Int] = Array(1, 3, 4, 6, 9, 10)

scala> List.concat(a,b)
val res67: List[Int] = List(1, 3, 4, 6, 9, 10, 1, 3, 5, 7, 9)

scala> a.intersect(b)
val res68: List[Int] = List(1, 3, 9)

scala> a.diff(b)
val res69: List[Int] = List(4, 6, 10)

scala> a(1)
val res70: Int = 3
```

图 3-36 列表的常用操作方法示例

除了以上可变和不可变列表共同的操作方法，还有一些操作方法只适用于可变列表，如表 3-5 所示。

表 3-5　可变列表的常用操作方法

方法（对列表 a 和 b 进行操作）	描　　述
a+= 元素	为列表添加元素
a++=b	追加一个列表
a-= 元素	删除列表中的元素
a.toList	将可变列表转换为不可变列表

在 scala 命令行中输入以下命令，对两个可变列表进行表 3-5 中的操作，结果如图 3-37 所示。

```
import scala.collection.mutable.ListBuffer
val a = ListBuffer(1,3,5,7,9)
val b = ListBuffer(2,3,4,7,8)
a += 3
a++=b
a-=1
a.toList
```

```
scala> import scala.collection.mutable.ListBuffer
import scala.collection.mutable.ListBuffer

scala> val a = ListBuffer(1,3,5,7,9)
val a: scala.collection.mutable.ListBuffer[Int] = ListBuffer(1, 3, 5, 7, 9)

scala> val b = ListBuffer(2,3,4,7,8)
val b: scala.collection.mutable.ListBuffer[Int] = ListBuffer(2, 3, 4, 7, 8)

scala> a += 3
val res71: a.type = ListBuffer(1, 3, 5, 7, 9, 3)

scala> a++=b
val res72: a.type = ListBuffer(1, 3, 5, 7, 9, 3, 2, 3, 4, 7, 8)

scala> a-=1
val res73: a.type = ListBuffer(3, 5, 7, 9, 3, 2, 3, 4, 7, 8)

scala> a.toList
val res74: List[Int] = List(3, 5, 7, 9, 3, 2, 3, 4, 7, 8)
```

图 3-37　可变列表的常用操作方法

由操作结果可知，虽然可变和不可变列表都有添加、删除、合并等操作，但是两者有一个非常大的差别，即对不可变列表进行操作，会产生一个新的列表，原来的列表并没有改变；而对可变列表进行操作，改变的是该列表本身。

对于不可变列表，还可以使用":::"进行两个列表的合并，在 scala 命令行中输入以下命令，合并两个不可变列表，结果如图 3-38 所示。

```
val a = List(1,3,4,6,9,10)
val b = List(7,3,3,6,8,10)
a:::b
```

```
scala> val a = List(1,3,4,6,9,10)
val a: List[Int] = List(1, 3, 4, 6, 9, 10)

scala> val b = List(7,3,3,6,8,10)
val b: List[Int] = List(7, 3, 3, 6, 8, 10)

scala> a:::b
val res75: List[Int] = List(1, 3, 4, 6, 9, 10, 7, 3, 3, 6, 8, 10)
```

图 3-38　使用 ":::" 进行两个列表的合并

2. 集合

集合（Set）是无序、不可重复的，分为可变的和不可变的。Scala 默认创建的是不可变集合，若要创建可变集合，则需要先导入依赖包。

（1）定义集合。

定义不可变集合的语法格式如下：

val 集合名 =Set[类型](值 1, 值 2, 值 3...)

例如，在 scala 命令行中输入以下命令，定义不可变集合，结果如图 3-39 所示。

val a = Set(9, 2, 5, 4, 8)

```
scala> val a = Set(9, 2, 5, 4, 8)
val a: scala.collection.immutable.Set[Int] = HashSet(5, 9, 2, 8, 4)
```

图 3-39　定义不可变集合

定义可变集合的语法与定义不可变集合的语法一致，但在定义之前需要先导入 "scala.collection.mutable.Set" 依赖包。

在 scala 命令行中输入以下命令，定义可变集合，结果如图 3-40 所示。

import scala.collection.mutable.Set
val b = Set(1, 10, 9, 2, 5, 4, 8)

```
scala> import scala.collection.mutable.Set
import scala.collection.mutable.Set

scala> val b = Set(1, 10, 9, 2, 5, 4, 8)
val b: scala.collection.mutable.Set[Int] = HashSet(1, 2, 4, 5, 8, 9, 10)
```

图 3-40　定义可变集合

（2）集合的常用操作方法。

可变和不可变集合的常用操作方法如表 3-6 所示。

表 3-6　可变和不可变集合的常用操作方法

方法（对集合 a 和 b 进行操作）	描述
a.drop(num)	删除从左边开始的 num 个元素
a.contains(元素)	判断集合是否包含此元素
a.size	集合的大小
a++b	合并集合

在 scala 命令行中输入以下命令，对不可变集合 a 和可变集合 b 执行表 3-6 中的操作，

结果如图 3-41 所示。

a.drop(2)
a.contains(1)
a.size
a++b
b.drop(2)
b.contains(1)
b.size

```
scala> a.drop(2)
val res0: scala.collection.immutable.Set[Int] = HashSet(2, 8, 4)

scala> a.contains(1)
val res1: Boolean = false

scala> a.size
val res2: Int = 5

scala> a++b
val res3: scala.collection.immutable.Set[Int] = HashSet(5, 10, 1, 9, 2, 8, 4)

scala> b.drop(2)
val res4: scala.collection.mutable.Set[Int] = HashSet(8, 9, 10, 4, 5)

scala> b.contains(1)
val res5: Boolean = true

scala> b.size
val res6: Int = 7
```

图 3-41　集合的常用操作

不可变集合的常用操作方法如表 3-7 所示。

表 3-7　不可变集合的常用操作方法

方法（对集合 a 进行操作）	描　　述
a + 元素	添加元素
a - 元素	删除元素

在 scala 命令行中输入以下命令，对不可变集合 a 执行表 3-7 中的操作，结果如图 3-42 所示。

a + 3
a - 3

```
scala> a+3
val res7: scala.collection.immutable.Set[Int] = HashSet(5, 9, 2, 3, 8, 4)

scala> a-3
val res8: scala.collection.immutable.Set[Int] = HashSet(5, 9, 2, 8, 4)
```

图 3-42　不可变集合的常用操作

在 scala 命令行中输入以下命令，对可变集合 b 执行"+"操作。由图 3-43 所示的报错结果可知，"+"不适用于为可变集合添加元素。

b + 3

大数据分析及应用项目教程 (Spark SQL)

```
scala> b + 3
        ^
        warning: method + in trait SetOps is deprecate
d (since 2.13.0): Consider requiring an immutable Set
 or fall back to Set.union
val res11: scala.collection.mutable.Set[Int] = HashSe
t(1, 2, 3, 4, 5, 8, 9, 10)
```

图 3-43　对可变集合 b 执行 "+" 操作报错

可变集合的常用操作方法如表 3-8 所示。

表 3-8　可变集合的常用操作方法

方法（对可变集合 b 进行操作）	描　　述
b.add(元素)	添加元素
b.remove(元素)	删除元素
b+=元素	添加元素
b-=元素	删除元素

在 scala 命令行中输入以下命令，对可变集合 b 执行表 3-8 中的操作，结果如图 3-44 所示。

b.add(3)
b.remove(3)
b+=3
b-=3

```
scala> b.add(3)
val res0: Boolean = true

scala> b
val res1: scala.collection.mutable.Set[Int] = HashSet(1, 2, 3, 4, 5, 8, 9, 10)

scala> b.remove(3)
val res2: Boolean = true

scala> b
val res3: scala.collection.mutable.Set[Int] = HashSet(1, 2, 4, 5, 8, 9, 10)

scala> b+=3
val res4: b.type = HashSet(1, 2, 3, 4, 5, 8, 9, 10)

scala> b-=3
val res5: b.type = HashSet(1, 2, 4, 5, 8, 9, 10)
```

图 3-44　可变集合的常用操作方法

集合与列表一样，在对不可变集合进行操作时会产生一个新的集合，原来的集合并没有改变。对可变集合进行操作，改变的是该集合本身。

3. 元组

元组（Tuple）是不同类型值的集合，它可以将不同类型的值放在一个变量中进行存储。定义元组的语法格式如下：

val tuple=(元素 1, 元素 2, 元素 3…)

在 scala 命令行中输入以下命令，定义一个元组，结果如图 3-45 所示。

val mytuple=(" 李雷 ",20)

```
scala> val mytuple=("李雷",20)
val mytuple: (String, Int) = (李雷,20)
```

图 3-45　定义一个元组（1）

定义元组的语法也可以写成以下格式：

val t=new Tuplen(元素 1, 元素 2, 元素 3, ..., 元素 n)

其中 n 表示元组中有 n 个元素。在 scala 命令行中输入以下命令，以第二种格式定义一个元组，结果如图 3-46 所示。

val t=new Tuple2(" 李雷 ",20)

```
scala> val t=new Tuple2("李雷",20)
val t: (String, Int) = (李雷,20)
```

图 3-46　定义一个元组（2）

当需要访问元组中某个元素的值时，可以通过"元组名 _ 元素索引"进行访问。在 scala 命令行中输入以下命令，访问元组中的元素，结果如图 3-47 所示。

println(" 我的名字是 "+t._1)
println(" 我的年龄是 "+t._2)

```
scala> println("我的名字是"+t._1)
我的名字是李雷

scala> println("我的年龄是"+t._2)
我的年龄是20
```

图 3-47　访问元组中的元素

4．映射

（1）映射的定义。

映射（Map）是一系列键 - 值对的集合。映射中的键都是唯一的，并且可以通过键获取值。Scala 默认创建的是不可变映射，其定义方式如下：

val 映射名 :Map[数据类型 , 数据类型] = Map(键 1 -> 值 , 键 2 -> 值 ...)
val 映射名 =Map((键 1, 值),(键 2, 值),(键 3, 值)...)

在 Scala 命令行中输入以下命令，用以上两种方法分别定义一个不可变映射，结果如图 3-48 所示。

val major=Map("101"->"bigdata","102"->"computer","103"->"electrical")
val major=Map((101,"bigdata"),(102,"computer"),(103,"electrical"))

```
scala> val major=Map("101"->"bigdata","102"->"computer","10
3"->"electrical")
val major: scala.collection.immutable.Map[String,String] =
Map(101 -> bigdata, 102 -> computer, 103 -> electrical)

scala> val major=Map((101,"bigdata"),(102,"computer"),(103,
"electrical"))
val major: scala.collection.immutable.Map[Int,String] = Map
(101 -> bigdata, 102 -> computer, 103 -> electrical)
```

图 3-48　定义一个不可变映射

可变映射的定义需要先导入相关依赖包,在 scala 命令行中输入以下命令,定义一个可变映射,结果如图 3-49 所示。

```
import scala.collection.mutable
val age = mutable.Map("李雷" -> 20,"韩美美" -> 21)
```

```
scala> import scala.collection.mutable
import scala.collection.mutable

scala> val age = mutable.Map("李雷" -> 20, "韩美美" -> 21)
val age: scala.collection.mutable.Map[String,Int] = HashMap(韩美美 -> 21, 李雷 -> 20)
```

图 3-49　定义一个可变映射

(2) 映射的常用操作方法。

映射的常用操作方法与列表、集合的操作方法是相同的。此外,映射还有一些其他的操作方法,如表 3-9 所示。

表 3-9　映射的常用操作方法

方法(对映射 a 和 b 进行操作)	描述
a.keys	获取所有的 key
a.values	获取所有的 value
a(key)	使用 key 获取 value,如果 key 不存在则会报错
a.contains(key)	检查映射中是否存在指定的 key
a.isEmpty	检查映射是否为空
a++b	合并两个映射,如果 key 重复,则 ++ 后的映射会替换前边 key 对应的 value

在 scala 命令行中输入以下命令,对前面定义的映射 major 和 age 进行表 3-9 中的操作,结果如图 3-50 所示。

```
major.keys
major.values
major("101")
major.contains("101")
major.isEmpty
major++age
```

```
scala> major.keys
val res0: Iterable[String] = Set(101, 102, 103)

scala> major.values
val res1: Iterable[String] = Iterable(bigdata, computer, electrical)

scala> major("101")
val res2: String = bigdata

scala> major.contains("101")
val res3: Boolean = true

scala> major.isEmpty
val res4: Boolean = false

scala> major++age
val res5: scala.collection.immutable.Map[String,Any] = HashMap(韩美美 -> 21, 李雷 -> 20, 103 -> electrical, 102 -> computer, 101 -> bigdata)
```

图 3-50　映射的常用操作方法

5. 组合器

组合器在实际操作中非常实用,包括 map、filter、foreach、groupBy、flatten、drop、zip 等,常用的组合器及其作用如表 3-10 所示。组合器可以对集合中的每个元素进行指定的操作,并将结果形成集合输出。

Scala 函数组合器

表 3-10 常用的组合器及其作用

组合器	描 述
map	将某个函数应用到集合中的所有元素中,并将结果形成集合输出
filter	指定条件,对集合中的元素进行过滤
foreach	对集合中的每个元素进行作用,但是没有返回值
groupBy	对集合中的元素进行分组操作,得到一个映射
flatten	把嵌套的结构展开
drop	去掉集合前面的 n 个元素
zip	将两个集合结合在一起

在 scala 命令行中输入以下命令,定义两个列表,结果如图 3-51 所示。

val a=List(2,3,7,5,9,1,6)
val b=List("a","b","c")

```
scala> val a=List(2,3,7,5,9,1,6)
val a: List[Int] = List(2, 3, 7, 5, 9, 1, 6)

scala> val b=List("a","b","c")
val b: List[String] = List(a, b, c)
```

图 3-51 定义两个列表

(1) map 组合器。

使用 map 组合器对列表 a 中的所有元素进行乘 2 计算,结果如图 3-52 所示。

a.map(x=>2*x)

```
scala> a.map(x=>2*x)
val res8: List[Int] = List(4, 6, 14, 10, 18, 2, 12)
```

图 3-52 map 组合器举例

(2) filter 组合器。

使用 filter 组合器对列表 a 中的所有元素进行过滤,输出大于 5 的元素,结果如图 3-53 所示。

a.filter(x=>x>5)

```
scala> a.filter(x=>x>5)
val res10: List[Int] = List(7, 9, 6)
```

图 3-53 filter 组合器举例

(3) foreach 组合器。

使用 foreach 组合器对列表 a 中的所有元素进行平方操作并打印输出,结果如图 3-54

所示。

a.foreach(x=>println(x*x))

```
scala> a.foreach(x=>println(x*x))
4
9
49
25
81
1
36
```

图 3-54　foreach 组合器举例

（4）groupBy 组合器。

使用 groupBy 组合器对列表 a 中的所有元素进行分组，分组条件是大于 5 的元素为一组，小于或等于 5 的元素为一组，结果如图 3-55 所示。

a.groupBy(x=>x>5)

```
scala> a.groupBy(x=>x>5)
val res13: scala.collection.immutable.Map[Boolean,List[Int]]
 = HashMap(false -> List(2, 3, 5, 1), true -> List(7, 9, 6))
```

图 3-55　groupBy 组合器举例

由图 3-55 可知，输出结果是一个映射，且符合 groupBy 分组条件的 List 值对应的键是 true，不符合 groupBy 分组条件的 List 值对应的键是 false。

（5）flatten 组合器。

定义一个二维列表，使用 flatten 组合器将其结构展开，结果如图 3-56 所示。

val c=List(List(1,6,4),List(2,5,8))
c.flatten

```
scala> val c=List(List(1,6,4),List(2,5,8))
val c: List[List[Int]] = List(List(1, 6, 4), List(2, 5, 8))

scala> c.flatten
val res15: List[Int] = List(1, 6, 4, 2, 5, 8)
```

图 3-56　flatten 组合器举例

（6）drop 组合器。

使用 drop 组合器去掉列表 a 前面的两个元素，结果如图 3-57 所示。

a.drop(2)

```
scala> a.drop(2)
val res16: List[Int] = List(7, 5, 9, 1, 6)
```

图 3-57　drop 组合器举例

（7）zip 组合器。

使用 zip 组合器将 a、b 两个列表结合在一起，结果如图 3-58 所示。

a.zip(b)

```
scala> a.zip(b)
val res18: List[(Int, String)] = List((2,a), (3,b), (7,c))
```

图 3-58　zip 组合器举例

由图 3-58 可知，在使用 zip 组合器时，最后产生的集合中元素的个数与两个列表中元素个数较少的那个相同。

二、编程分析实现

以班级为单位整理学生信息，可以先将数据保存到列表中，然后使用 groupBy 组合器以班级为条件进行分组。此任务将截取部分原始数据进行操作。

在 scala 命令行中输入以下命令，创建列表保存数据，结果如图 3-59 所示。

```
val data=List(
    "21103,严林石,男,21级,大数据1班,89,97,91",
    "21108,庄伟康,男,21级,大数据1班,67,94,87",
    "21208,陈心雨,女,21级,大数据2班,84,92,91",
    "21107,邹和俊,男,21级,大数据1班,77,83,88")
```

```
scala> val data=List(
     |     "21103,严林石,男,21级,大数据1班,89,97,91",
     |     "21108,庄伟康,男,21级,大数据1班,67,94,87",
     |     "21208,陈心雨,女,21级,大数据2班,84,92,91",
     |     "21107,邹和俊,男,21级,大数据1班,77,83,88")
val data: List[String] = List(21103,严林石,男,21级,大数据1班,89,97,91, 21108,庄伟康,男,21级,大数据1班,67,94,87, 21208,陈心雨,女,21级,大数据2班,84,92,91, 21107,邹和俊,男,21级,大数据1班,77,83,88)
```

图 3-59　创建列表保存数据

使用 groupBy 组合器进行分组，分组条件是班级。因此需要选取班级列进行操作，使用 "," 对列表中的每条数据进行分隔，班级数据是在第 4 列。在 scala 命令行中输入以下命令，结果如图 3-60 所示。

```
data.groupBy(x=>x.split(",")(4))
```

```
scala> data.groupBy(x=>x.split(",")(4))
val res20: scala.collection.immutable.Map[String,List[String]]
 = HashMap(大数据2班 -> List(21208,陈心雨,女,21级,大数据2班,84,92,91), 大数据1班 -> List(21103,严林石,男,21级,大数据1班,89,97,91, 21108,庄伟康,男,21级,大数据1班,67,94,87, 21107,邹和俊,男,21级,大数据1班,77,83,88))
```

图 3-60　使用 groupBy 组合器进行分组

由图 3-60 可知，键为"大数据 2 班"的值有 1 条，即在原列表中有一位同学是大数据 2 班的；键为"大数据 1 班"的值有 3 条，即在原列表中有 3 位同学是大数据 1 班的。由此实现了以班级为单位整理学生信息。

任务 3.2　学生基本情况分析

情境导入

在掌握了"任务 3.1 班级基本情况分析"中 Scala 的基本知识之后，李雷同学提出了新的疑问。与其他编程语言相比，Scala 作为一门纯粹的面向对象的语言，它的类和对象有什么特点？定义和使用方法如何？接下来通过对 Scala 类和对象、模式匹配等知识的讲解，对学生特长情况与成绩情况进行分析。

学习目标和要求

知识与技能目标

1. 了解 Scala 类和对象、模式匹配的基本概念。
2. 掌握 Scala 类和对象、模式匹配的定义及操作方法。
3. 综合应用 Scala 基础知识，编写应用程序进行数据分析。

素质目标

1. 提升严谨细致的编程逻辑思维能力。
2. 提升分析问题、解决问题的能力。

3.2.1　学生特长情况分析

Scala 类和对象

一、理论基础

1. 类和对象

类和对象是 Scala 语言的两个重要概念。类是对象的抽象，而对象是类的具体实例。类的定义方式如下：

```
Class 类名称 ( 参数列表 ) {
  // 定义类的字段和方法
}
```

在 scala 命令行中输入以下命令，定义一个 Physique 类，记录学生的身高和体重，结果如图 3-61 所示。（注意：类名的首字母要大写。）

```
class Physique(vname:String, vheight:Int) {
  var name=vname
  var height=vheight
```

```
  def add(dheight:Int): Unit ={
  height=vheight+dheight
  println("myname: "+vname+", myheight: "+height+"cm")
  }
}
new Physique("李雷",170).add(1)
```

```
scala> class Physique(vname:String, vheight:Int) {
     | var name=vname
     | var height=vheight
     | def add(dheight:Int): Unit ={
     | height=vheight+dheight
     | println("myname: "+vname+", myheight: "+height+"cm")
     | }
     | }
class Physique

scala> new Physique("李雷",170).add(1)
myname: 李雷, myheight: 171cm
```

图 3-61　定义一个 Physique 类

在图 3-61 中，定义了一个名为 Physique 的类，该类包含 name 和 age 两个参数；该类中定义了一个 add() 方法。在完成定义类之后，使用 new() 方法将类实例化。

对 Physique 类追加"学生体重"属性，同时保留原来的 Physique 类，此时需要用到继承。继承是面向对象开发语言中的一个概念，它可以使子类具有父类的各种属性和方法，而不需要再次编写相同的代码，可以有效复用代码；或者重新定义、追加类的属性和方法等。但是 Scala 是单继承，只能继承一个父类。继承类需要用到关键字 extends。继承的语法格式如下：

class 子类名 extends 父类名 {类体}

在 scala 命令行中输入以下命令，定义一个 Physiquem 类，并继承 Physique 类，结果如图 3-62 所示。

```
class Physiquem(vname:String,vheight:Int,vweight:Int) extends Physique(vname, vheight){
  var weight=vweight
  def add(dheight:Int, dweight:Int): Unit ={
  height=vheight+dheight
  weight=vweight+dweight
  println("myname: "+vname+", myheight: "+height+"cm, myweight: "+weight+"kg")
  }
}
new Physiquem("李雷",170,65).add(1,1)
```

```
scala> class Physiquem(vname:String,vheight:Int,vweight:Int)
extends Physique(vname, vheight){
     | var weight=vweight
     | def add(dheight:Int, dweight:Int): Unit ={
     | height=vheight+dheight
     | weight=vweight+dweight
     | println("myname: "+vname+", myheight: "+height+"cm, my
weight: "+weight+"kg")
     | }
     | }
class Physiquem

scala> new Physiquem("李雷",170,65).add(1,1)
myname: 李雷, myheight: 171cm, myweight: 66kg
```

图 3-62　Physiquem 类继承 Physique 类

注意，子类在继承父类中非抽象的方法（已经实现的方法）时必须使用 override 修饰符。

2. 单例对象和伴生对象

（1）单例对象。

单例对象是一种特殊的类，有且只有一个实例。单例对象类似 Java 的静态类，它的成员、方法默认都是静态的。单例对象与类的区别是单例对象不能带参数。定义一个单例对象的语法格式如下：

object 单例对象名 { }

在 scala 命令行中输入以下命令，定义一个单例对象，结果如图 3-63 所示。

```
object Personid{
var ID = 0
def idnum:Int={
  ID += 1
  ID
}}
```

图 3-63　定义一个单例对象

（2）伴生对象。

若在同一个代码文件内部，同时出现了 class A 和 object A，即类名和单例对象名完全相同，那么它们互为伴生关系，类称为对象的伴生类，对象称为类的伴生对象。伴生关系中的对象和类彼此可以访问对方的私有成员。

在 scala 命令行中输入以下命令，定义伴生类和伴生对象，并相互访问，结果如图 3-64 所示。

```
// 定义伴生类
class Student(var name:String){
  // 访问伴生对象的私有成员
  def info() = println(name + " 编号 "+Student.no)
}
// 定义伴生对象
object Student {
  private var no:Int = 0
  def sno()={
    no += 1
    no
  }
```

}
// 调用单例对象
println(" 单例对象,编号: " + Student.sno())
// 实例化伴生类
val obj = new Student(" 李雷 ")
obj.info()

```
scala> //定义伴生类
     | class Student(var name:String){
     | //访问伴生对象的私有成员
     | def info() = println(name + "编号"+Student.no)
     | }
     | //定义伴生对象
     | object Student {
     | private var no:Int = 0
     | def sno()={
     | no += 1
     | no
     | }
     | }
     | //调用单例对象
     | println("单例对象,编号: " + Student.sno())
     | //实例化伴生类
     | val obj = new Student("李雷")
     | obj.info()
单例对象,编号: 1
李雷编号1
class Student
object Student
val obj: Student = Student@7489bca4
```

图 3-64　定义一个伴生对象

在图 3-64 中,定义了一个伴生类 Student,在类中定义了一个 info() 方法;还定义了此伴生类的伴生对象,伴生类访问了伴生对象的 sno() 方法。

3. 模式匹配

模式匹配（Pattern Matching）是 Scala 语言的一个十分强大的特性。模式匹配的基本语法如下:

Scala 模式匹配

```
x match {
  case pattern1 => do Something
  case pattern2 => do others
  ...}
```

其中 x 为变量,后面紧跟 match 关键字声明。代码块中的每个分支均采用 case 关键字进行声明。在进行模式匹配时,从第一个 case 分支开始,如果匹配成功,则执行"=>"符号后面的逻辑代码;如果匹配不成功,则继续对下一个分支进行匹配。如果所有的 case 分支都不匹配,则执行 case _ 分支。一旦某个 case 匹配成功,则剩下的将不再继续匹配。

在 scala 命令行中输入以下命令,进行模式匹配,结果如图 3-65 所示。

```
def matchstu(x: Int): String = x match {
    case 21103 => " 严林石 "
    case 21102 => " 王西茜 "
```

```
    case 21201 => "吴倩瑶"
    case _ => "otherstu"
}
```
println(matchstu(21103))
println(matchstu(66666))

```
scala> def matchstu(x: Int): String = x match {
     |     case 21103 => "严林石"
     |     case 21102 => "王西茜"
     |     case 21201 => "吴倩瑶"
     |     case _ => "otherstu"
     | }
def matchstu(x: Int): String

scala> println(matchstu(21103))
严林石

scala> println(matchstu(66666))
otherstu
```

图 3-65　模式匹配举例

由图 3-65 可知，当匹配到对应的学号时，则输出学生的名字；当没有匹配到任何学号时，则输出 otherstu。

在定义类的时候，如果使用了 case 关键字，则这个类称为样例类（Match Class）。样例类是为了模式匹配而优化的类。

在 scala 命令行中输入以下命令，定义样例类，并在模式匹配中调用该样例类，结果如图 3-66 所示。

```
// 定义 Identity 类，最简单的类的定义方法是只有一个 class 关键字和一个表示类名的标识符
class Identity
// 定义样例类 Student
case class Student(name: String, age: Int, stuNo: String) extends Identity
// 定义样例类 Teacher
case class Teacher(name: String, age: Int, teaNo: String) extends Identity
// 定义样例类 Otherbody
case class otherbody(name: String) extends Identity
// 创建对象 p，样例类会自动生成 apply() 方法，在创建对象时无须使用 new() 方法
val p: Identity = Student("lisi", 20, "101")
// 进行 Match Case 模式匹配
p match {
case Student(name, age, stuNo) => println(s"学生：$name, $age, $stuNo")
case Teacher(name, age, teaNo) => println(s"老师：$name, $age, $teaNo")
case Otherbody(name) => println(s"其他人：$name")
}
```

第 3 章 学生信息处理分析

```
scala> class Identity
     | //定义样例类Student
     | case class Student(name: String, age: Int, stuNo: String) extends Identity
     | //定义样例类Teacher
     | case class Teacher(name: String, age: Int, teaNo: String) extends Identity
     | //定义样例类Otherbody
     | case class otherbody(name: String) extends Identity
     | //创建对象p，case class会自动生成apply()方法，在创建对象时无须使用new()方法
     | val p: Identity = Student("lisi", 20, "101")
     | //进行Match Case模式匹配
     | p match {
     | case Student(name, age, stuNo) => println(s"学生：$name, $age, $stuNo")
     | case Teacher(name, age, teaNo) => println(s"老师：$name, $age, $teaNo")
     | case otherbody(name) => println(s"其他人：$name")
     | }
学生：lisi, 20, 101
class Identity
class Student
class Teacher
class otherbody
val p: Identity = Student(lisi,20,101)
```

图 3-66　样例类举例

二、编程分析实现

学生特长情况数据（见"speciality.txt"文件）包括 3 个字段，分别是学生学号、姓名、特长类别。现要获取具有某一类特长的学生的情况，可以先定义一个 object 单例对象，在此对象中读取数据文件，并转换成列表；接着遍历列表，搜索具有该类特长的学生的数据并打印。

案例实现——学生特长情况获取

此处涉及 Scala 读取文件的方法。可以使用 Scala 中的 Source 单例对象来读取文件。Source 单例对象提供了一些非常便捷的方法，可以使开发者快速地从指定数据源（如文本文件、URL 地址等）中获取数据。在使用 Source 单例对象之前，需要先导入"import scala.io.Source"依赖包，其语法格式如下：

Source.fromFile(文件存储地址)

在 scala 命令行中输入以下命令，获取音乐类（musical）特长生的情况，结果如图 3-67 所示。

```
import scala.io.Source
object Speciality{
// 从本地目录中读取数据文件
val source=Source.fromFile("/root/data/Chapter3/speciality.txt")
// 按行进行数据读取
val lines=source.getLines()
// 将数据转换为列表
val list=lines.toList
// 定义函数，判断每一行是否包含指定的字符串
def checkclassify(classify:String): Unit ={
    for(line<- list;if line.contains(classify)){println(line)}
}
```

}
Speciality checkclassify "musical"

```
scala> import scala.io.Source
     | object Speciality{
     | val source=Source.fromFile("/root/data/Chapter3/speciality.txt")
     | val lines=source.getLines()
     | val list=lines.toList
     | def checkclassify(classify:String): Unit ={
     | for(line<- list;if line.contains(classify)){println(line)}
     | }
     | }
import scala.io.Source
object Speciality

scala> Speciality checkclassify "musical"
21203,严林石生,musical
21206,沈高翔,musical
21210,叶嘉乐,musical
21216,江明晓,musical
21222,陈韩旭,musical
21229,陈可可,musical
21233,李柔娴,musical
21238,陈鑫涛,musical
21203,郑翔天,musical
21207,杨溢,musical
21213,徐漓,musical
21219,王国龙,musical
21220,应雨晨,musical
21228,李微,musical
21234,徐驰,musical
21237,史华锋,musical
21243,郑晔超,musical
```

图 3-67　音乐类特长生的情况

由图 3-67 可知，在定义完成 Speciality 单例对象之后，调用了单例对象中的方法，查询音乐类特长生的情况，有 17 位学生的特长是音乐类。

3.2.2　学生成绩情况分析

图 3-1 所示的原始数据包含的数据字段如表 3-11 所示。

表 3-11　"data.txt" 原始数据字段

字　　段	含　　义
stuid	学号
name	姓名
sex	性别
grade	年级
class	班级
Chinese	语文成绩
Math	数学成绩
English	英语成绩

第 3 章　学生信息处理分析

现要求使用 Scala 语言进行函数式编程，综合前面所学知识，统计并分析学生成绩，获取各门课程的平均成绩和及格人数（分数大于 60 分），以及每个学生的总成绩。

（1）导入依赖包，读取数据文件。

将数据文件"data.txt"上传到本地"/root/data/Chapter3/"目录下，并通过 Source 单例对象读取。

```
import scala.io.Source
val inputfile = Source.fromFile("/root/data/Chapter3/data.txt")
```

（2）将数据转换为 List 列表。

由于在后续的分析操作中会对数据进行多次遍历，因此需要将其转换为列表。同时观察原始数据，每一行中的各项数据均以","分隔。

```
val originaldata = inputfile.getLines().map{_.split(",")}.toList
```

（3）创建课程名称列表。

原始数据文件中的第一行没有数据字段名，因此需要自行创建一个内容为课程名称的列表，用以匹配并显示每门课的平均分等信息。

```
val coursename =List(" 语文 "," 数学 "," 英语 ")
```

（4）定义函数，计算各门课程的平均成绩和及格人数。

calculation1() 函数传入的参数是一个列表。3 门课程的成绩数据在列表的第 5、6、7 列，使用 for 循环，对这 3 列数据进行处理。使用 map() 方法将这 3 列数据的数据类型转换为 Double 数据类型。使用 sum() 函数分别对这 3 列数据求和，并除以列表的长度（每列数据的个数），就能获得每门课程的平均成绩。使用 filter() 方法对每列数据进行过滤，获得分数大于 60 的元素。使用 length() 方法求出过滤后列表的长度，即及格的人数。最终生成每门课程的二元组，包括平均成绩和及格人数。

```
def calculation1 (lines:List[Array[String]])={
  for (i<-5 to 7)yield{
    val temp = lines map{
    elem=>elem(i).toDouble
    }
    (temp.sum/lines.length,temp.filter(s=>s>=60.0).length)
}}
```

（5）定义函数，统计每个学生的总成绩。

calculation2() 函数传入的参数是一个列表。列表中每一行表示一个学生的信息，因为列表从 0 开始计数，所以学生的总数可以用 lines.length-1 表示。通过 for 循环将每个学生的数据处理成一个二元组，包括学生的学号和总成绩。

在 for 循环中，首先对每一行使用 toList() 方法，将其转换为列表，用 head() 方法获取学生的学号（第一个元素）。然后对每一行使用 drop(5) 方法，删除前面的信息，只保留成绩信息。最后使用 toArray() 方法将成绩信息转换成数组，将每个元素的数据类型都转换成 Double 数据类型，求和获得每个学生所有课程的总成绩。

```
def calculation2 (lines:List[Array[String]])={
```

```
  for (i<-0 to lines.length-1)yield{
    (lines(i).toList.head,lines(i).drop(5).toArray.map(s=>s.toDouble).sum)
  }
}
```

（6）定义函数，用于输出 calculation1() 函数的计算结果。

calculation1() 函数获得的结果是每门课程的平均分和及格人数，但是没有对应的课程名称。所以在输出之前需要将课程名容器和结果容器合并为二元组容器，并通过模式匹配进行输出。输出格式 "-6s" 中的 "-" 表示左对齐，6 表示字段宽度，s 表示字符串格式；"5.2f" 中小数点后的 2 表示保留两位小数，f 表示单精度浮点型数据。

```
def printresult (theresult:Seq[(Double,Int)]): Unit ={
  (coursename zip theresult)foreach{
    case (name,result)=>
    println(f"${name+":"}%-6s${result._1}%5.2f${result._2}%10f")}
}
```

（7）综合上述函数与方法，进行调用、统计、分析与输出操作。

在 scala 命令行中输入以下命令，获取各门课程的平均成绩与及格人数，结果如图 3-68 所示。

```
val result1 = calculation1(originaldata)
println("course   average    >60")
printresult(result1)
```

```
scala> val result1 = calculation1(originaldata)
     | println("course   average    >60")
     | printresult(result1)
course   average    >60
语文：    81.53       19.00
数学：    85.58       19.00
英语：    84.58       19.00
val result1: IndexedSeq[(Double, Int)] = Vector((81.
52631578947368,19), (85.57894736842105,19), (84.5789
4736842105,19))
```

图 3-68　各门课程的平均成绩与及格人数

在 scala 命令行中输入以下命令，每个学生的总成绩如图 3-69 所示。

```
val result2 = calculation2(originaldata)
println("Id       sum")
for (i<-0 to result2.length-1){
  println(f"${result2(i)._1}%-10s${result2(i)._2}%5.2f")
}
```

第 3 章　学生信息处理分析

```
scala> val result2 = calculation2(originaldata)
     | println("Id          sum")
     | for (i<-0 to result2.length-1){
     | println(f"${result2(i)._1}%-10s${result2(i)._2}%5.2f")
     | }
Id          sum
21103       277.00
21203       272.00
21102       257.00
21201       244.00
21104       253.00
21206       231.00
21105       241.00
21106       254.00
21110       229.00
21108       248.00
21208       267.00
21111       243.00
21109       245.00
21204       257.00
21202       276.00
21107       248.00
21205       244.00
21101       256.00
21207       240.00
val result2: IndexedSeq[(String, Double)] = Vector((21103,277.
0), (21203,272.0), (21102,257.0), (21201,244.0), (21104,253.0)
, (21206,231.0), (21105,241.0), (21106,254.0), (21110,229.0),
(21108,248.0), (21208,267.0), (21111,243.0), (21109,245.0), (2
1204,257.0), (21202,276.0), (21107,248.0), (21205,244.0), (211
01,256.0), (21207,240.0))
```

图 3-69　每个学生的总成绩

脑图小结

本章详细介绍了 Scala 的一些基本知识和操作方法，包括函数、表达式、判断与循环、数据结构、类和对象等，并完成了"班级基本情况分析"和"学生基本情况分析"任务。通过以下脑图小结，助力学习者掌握并巩固相关知识。

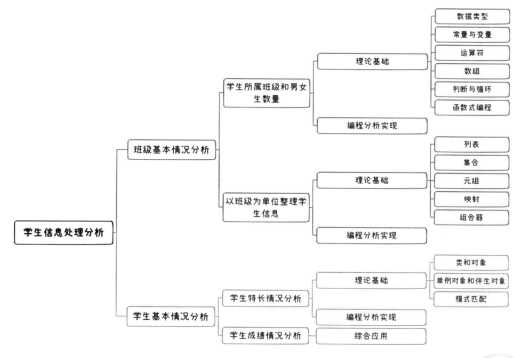

章节练习

1. 选择题

（1）以下选项不属于 Scala 数据类型的是（　　）。

　　A．Char　　　　　　B．Int　　　　　C．LongLong　　　　　D．Float

（2）以下选项属于 Scala 基本特性的有（　　）。

　　A．是一门类 Java 的多范式语言

　　B．是一门函数式语言，支持高阶函数

　　C．运行于 JVM 之上，并且兼容现有的 Java 程序

　　D．是一门纯粹的面向对象的语言

（3）以下关于 Scala 变量定义和赋值的命令中，错误的是（　　）。

　　A．val a=1

　　B．val a:String=4

　　C．var b:Int=2;b=4

　　D．var b="hello scala";b="12345"

（4）表达式 for(i <- 1 to 2; j <- 1 to 3; if i != j) {print((10 * i + j));print(" ")} 的输出结果是（　　）。

　　A．12 13 21 22 23

　　B．11 13 21 23

　　C．12 13 21 23

　　D．11 12 21 22

（5）以下用 Scala 语言定义的 List 中，不正确的是（　　）。

　　A．val list = List(4,2,0)

　　B．val list = List [String]('A','B','C')

　　C．val list = List [String]()

　　D．val list = List [Int](5,7,9)

（6）对于 Map（"bread" -> 2，"milk" -> 1).map(m => m._1 -> m._2 + 2) 的结果，正确的是（　　）。

　　A．Map("bread" -> 4, "milk" -> 3)

　　B．Map("breadbread" -> 4, "milkmilk" -> 3)

　　C．Map("bread" -> 2, "milk" -> 1 ,"bread" -> 2, "milk" -> 1)

　　D．Map("breadbread" -> 2, "milkmilk" -> 1)

（7）以下关于 Scala 各种数据结构的说法中，正确的是（　　）。

　　A．集合（Set）中的所有元素都不能重复

　　B．列表（List）一旦被定义，其值就不能改变

C．Scala object 的对象都是单例静态的

D．映射（Map）是一系列键-值对的容器，在一个映射中，键和值都是唯一的

（8）下列方法中，可以正确计算数组长度的是（　　）。

　　A．count　　　　　　B．take　　　　　C．length　　　　　　D．tail

（9）以下关于类和它的伴生对象的说法中，错误的是（　　）。

　　A．类和它的伴生对象可以有不同的名称

　　B．类和它的伴生对象定义在同一个文件中

　　C．类有静态方法，伴生对象没有静态方法

　　D．类和它的伴生对象可以互相访问私有特性

（10）以下关于类和单例对象的区别中，正确的是（　　）。

　　A．单例对象和类均可以定义方法

　　B．单例对象不接受参数，而类可以

　　C．单例对象和类均可以继承

　　D．单例对象和类均可以定义私有属性

2．填空题

（1）Scala 语言的特性有：_____、_____、_____、_____、_____。

（2）在 Scala 中，声明变量的关键字是_____，声明常量的关键字是_____。

（3）在 Scala 中获取列表中首个元素的方法是_____，获取列表中除首个元素以外的剩余部分的方法是_____。

（4）有一个数组 val a = Array(1,2,3)，对其执行操作 val b = a.map(_*2)，b 的结果是_____。

（5）在 Scala 中，模式匹配的关键字是由_____和_____组成的。

3．操作题

在日常阅读一篇文章时，我们可以通过某一个字段出现的次数了解文章的重点、关键字，以理解作者想表达的思想。现有一份文本数据"dream.txt"，记录了一篇泰戈尔的诗——《梦想》。应用所学的 Scala 知识，统计诗中"dream"这个单词出现的次数。

第 4 章

房产大数据分析与探索

任务 4.1　某房产公司销售人员业绩分析

情境导入

员工业绩考核分析是企业工作总结的重要组成部分。员工可以从业绩考核分析中清楚地知道自己的工作应该达到何种标准，发现自己的长处和不足，培养工作的积极性。管理者也可以从中清楚地了解员工的工作情况，通过分析来帮助员工改进不足、提高业绩，从而促进企业的发展。

假设现有某房产公司员工某年的销售业绩数据，请帮助部门经理对其进行统计、分析，并获取相关信息。相关数据表（见"staff.txt""first-half-year.txt""second-half-year.txt"文件）的字段及其含义如表 4-1、表 4-2、表 4-3 所示。

表 4-1　员工信息表

员工编号	姓　　名
11020101	马文轩
11040204	黄焱
12080403	沈睿广
10040106	王向秋
21050309	李悦可
22010611	许辉
20100706	杨昆明
19080303	任晓燕
18090313	廖文敏

续表

员工编号	姓　　　名
19111001	尚梦菲
11012008	徐悠
11022203	刘聪
21110301	刘浩丽
21010404	李安安
15012511	王嘉勋

表 4-2　上半年业绩表

员工编号	时　　　间	房屋销售套数	房屋销售金额（万元）
11020101	上半年	12	3400
11040204	上半年	10	3500
12080403	上半年	14	4200
10040106	上半年	8	5010
21050309	上半年	15	6300
22010611	上半年	6	3000
20100706	上半年	11	4800
19080303	上半年	13	4300
18090313	上半年	15	6000
19111001	上半年	16	5900
11012008	上半年	4	800
11022203	上半年	8	2300
21110301	上半年	10	3000
21010404	上半年	18	6100
15012511	上半年	20	8000

表 4-3　下半年业绩表

员工编号	时　　　间	房屋销售套数	房屋销售金额（万元）
11020101	下半年	10	3701
11040204	下半年	13	4500
12080403	下半年	10	4100
10040106	下半年	12	4010
21050309	下半年	18	5300
22010611	下半年	14	3000
20100706	下半年	8	4200
19080303	下半年	10	4300
18090313	下半年	8	5600
19111001	下半年	12	5510
11012008	下半年	7	1600

续表

员工编号	时间	房屋销售套数	房屋销售金额（万元）
11022203	下半年	10	3300
21110301	下半年	8	3000
21010404	下半年	15	5800
15012511	下半年	17	7020

学习目标和要求

知识与技能目标

1. 掌握将数据构建为 RDD 的方法。
2. 能使用 RDD 的各种转换和行动操作对数据集进行处理分析。
3. 能对处理完成的数据进行存储操作。

素质目标

1. 具有使用数据思维解决问题的意识。
2. 具有敬业爱岗、认真负责的职业品格。

4.1.1 数据集处理

Spark 中最核心、最基础的概念是弹性分布式数据集（Resilient Distributed Datasets，RDD）。它是一种分布式的内存抽象，可以基于任何数据结构进行创建。RDD 模型的设计初衷主要是为了提升运行效率、支持多样计算种类的广泛性、提升容错性，它具备不可变、高容错、分区存储、持久、分区算法等特性。在利用 Spark SQL 进行数据操作之前，要先创建 RDD。接下来从对数据创建 RDD 开始，介绍弹性分布式数据集的相关操作。

创建 RDD 的方法有两种，一种是基于内存（集合）创建，另一种是从外部数据集中创建。

1. 基于内存（集合）创建 RDD

基于内存（集合）创建 RDD，主要用到两个方法：SparkContext 类中的 parallelize() 和 makeRDD()。

在 master 节点中输入以下命令，启动 Spark 集群和 spark-shell，启动成功的结果如图 4-1 所示。

```
[root@master ~]# start-spark-all.sh
[root@master ~]# spark-shell
```

```
[root@master ~]# start-spark-all.sh
starting org.apache.spark.deploy.master.Master, logging to /usr/local/src/
spark/logs/spark-root-org.apache.spark.deploy.master.Master-1-master.out
slave02: starting org.apache.spark.deploy.worker.Worker, logging to /usr/l
ocal/src/spark/logs/spark-root-org.apache.spark.deploy.worker.Worker-1-sla
ve02.out
slave01: starting org.apache.spark.deploy.worker.Worker, logging to /usr/l
ocal/src/spark/logs/spark-root-org.apache.spark.deploy.worker.Worker-1-sla
ve01.out
localhost: starting org.apache.spark.deploy.worker.Worker, logging to /usr
/local/src/spark/logs/spark-root-org.apache.spark.deploy.worker.Worker-1-m
aster.out
[root@master ~]# spark-shell
2022-10-11 10:23:28,640 WARN util.NativeCodeLoader: Unable to load native-
hadoop library for your platform... using builtin-java classes where appli
cable
Setting default log level to "WARN".
To adjust logging level use sc.setLogLevel(newLevel). For SparkR, use setL
ogLevel(newLevel).
2022-10-11 10:23:41,116 WARN util.Utils: Service 'SparkUI' could not bind
on port 4040. Attempting port 4041.
Spark context Web UI available at http://master:4041
Spark context available as 'sc' (master = local[*], app id = local-1665498
221462).
Spark session available as 'spark'.
Welcome to
      ____              __
     / __/__  ___ _____/ /__
    _\ \/ _ \/ _ `/ __/  '_/
   /___/ .__/\_,_/_/ /_/\_\   version 3.1.2
      /_/

Using Scala version 2.12.10 (Java HotSpot(TM) 64-Bit Server VM, Java 1.8.0
_281)
Type in expressions to have them evaluated.
Type :help for more information.

scala>
```

图 4-1　启动 Spark 集群和 Spark-shell

1）使用 parallelize() 方法创建 RDD

使用此方法需要输入两个参数，第一个参数用来接收一个集合；第二个参数是可选的，用于指定 RDD 的分区数。RDD 中的每个分区都对应一个 Spark 任务，如果不设置此参数，则一般默认为应用程序使用的 CPU 核数。

在 spark-shell 中输入以下命令，使用 parallelize() 方法对列表创建 RDD，结果如图 4-2 所示。

// 定义一个列表
val list = List(1,2,3,4,5,6)

// 创建一个 RDD
val rdd = sc.parallelize(list)

// 查看返回值
rdd.collect

// 查看此 RDD 的默认分区
rdd.partitions.size

// 设置此 RDD 的分区数为 2

```
val rdd = sc.parallelize(list,2)

// 查看设置分区后的结果
rdd.partitions.size
```

```
scala> val list = List(1,2,3,4,5,6)
list: List[Int] = List(1, 2, 3, 4, 5, 6)

scala> val rdd = sc.parallelize(list)
rdd: org.apache.spark.rdd.RDD[Int] = ParallelCollectionRDD[2] at parallelize at <console>:26

scala> rdd.collect
res3: Array[Int] = Array(1, 2, 3, 4, 5, 6)

scala> rdd.partitions.size
res4: Int = 1

scala> val rdd = sc.parallelize(list,2)
rdd: org.apache.spark.rdd.RDD[Int] = ParallelCollectionRDD[3] at parallelize at <console>:26

scala> rdd.partitions.size
res5: Int = 2
```

图 4-2　使用 parallelize() 方法对列表创建 RDD

由图 4-2 可知，如果在创建 RDD 时不设置分区数，则默认为该应用程序分配的 CPU 核数，此处是 1。当设置分区数为 2 时，可以看到查询结果中显示的分区数是 2。RDD 有多少个分区就对应有多少个任务。

2）使用 makeRDD() 方法创建 RDD

makeRDD() 方法和 parallelize() 方法类似，但它可以指定每个分区的首选位置。分区首选位置决定了这条数据首选放在哪个分区中。

在 spark-shell 中输入以下命令，使用 makeRDD() 方法对数组创建 RDD，结果如图 4-3 所示。

```
// 定义一个数组
val arr = Array("good good study","day day up")

// 创建一个 RDD
val arrRDD = sc.makeRDD(arr)

// 查看返回值
arrRDD.collect
```

```
scala> val arr = Array("good good study","day day up")
arr: Array[String] = Array(good good study, day day up)

scala> val arrRDD = sc.makeRDD(arr)
arrRDD: org.apache.spark.rdd.RDD[String] = ParallelCollectionRDD[2] at makeRDD at <console>:26

scala> arrRDD.collect
res2: Array[String] = Array(good good study, day day up)

scala>
```

图 4-3　使用 makeRDD() 方法对数组创建 RDD

2. 从外部数据集中创建 RDD

从外部数据集中创建 RDD 是指，通过读取一个放在文件系统中的数据来创建 RDD。该数据集可以来自本地文件，主要用于测试；也可以由 HDFS 文件获取，这是实际操作中最常用的方法，主要针对 HDFS 上存储的大数据进行离线批处理操作。

从外部数据集中创建的 RDD 支持读取多种类型的数据集，如文本文件、压缩文件等。通过调用 SparkContext 的 textFile() 方法读取数据集，调用命令如下：

```
textFile("/my/directory")
textFile("/my/directory/*.txt")
textFile("/my/directory/*.gz")
```

1）从 HDFS 上存储的 CSV 文件中创建 RDD

（1）启动 Hadoop 集群，将 CSV 文件上传至 HDFS 中。

在 master 节点中输入以下命令，启动 Hadoop 集群并查看集群各项服务是否正常启动，结果如图 4-4 所示，表示启动正常。

```
[root@master ~]# start-all.sh
Starting namenodes on [master]
Last login: Tue Oct 11 11:08:55 EDT 2022 on pts/1
Starting datanodes
Last login: Tue Oct 11 11:09:37 EDT 2022 on pts/1
Starting secondary namenodes [master]
Last login: Tue Oct 11 11:09:40 EDT 2022 on pts/1
Starting resourcemanager
Last login: Tue Oct 11 11:09:54 EDT 2022 on pts/1
Starting nodemanagers
Last login: Tue Oct 11 11:10:10 EDT 2022 on pts/1
[root@master ~]# jps
3809 SparkSubmit
7175 ResourceManager
6696 DataNode
4057 Worker
3979 Master
4107 SparkSubmit
6907 SecondaryNameNode
7661 Jps
7310 NodeManager
6559 NameNode
[root@master ~]#
```

图 4-4 启动 Hadoop 集群并查看集群各项服务情况

将书本资料数据文件夹 "Chapter4" 中的 "test.csv" 文件上传到本地 Linux 的 "/root/data/Chapter4" 目录下，如图 4-5 所示。若没有此文件目录，则输入以下命令：

[root@master ~]# mkdir -p /root/data/Chapter4

图 4-5 上传文件到本地 Linux 中

继续输入以下命令，将"test.csv"文件上传至 HDFS 的"/Chapter4"目录下，若没有此目录则新建一个，并查看文件是否上传成功，操作过程与结果如图 4-6 所示。

```
[root@master ~]# hdfs dfs -mkdir /Chapter4
[root@master ~]# hdfs dfs -put /root/data/Chapter4/test.csv /Chapter4
[root@master ~]# hdfs dfs -ls /Chapter4
```

```
[root@master ~]# hdfs dfs -mkdir /Chapter4
[root@master ~]# hdfs dfs -put /root/data/Chapter4/test.csv /Chapter4
2022-10-11 11:14:43,347 INFO sasl.SaslDataTransferClient: SASL encryption t
rust check: localHostTrusted = false, remoteHostTrusted = false
[root@master ~]# hdfs dfs -ls /Chapter4
Found 1 items
-rw-r--r--   3 root supergroup        115 2022-10-11 11:14 /Chapter4/test.c
sv
[root@master ~]#
```

图 4-6　上传文件至 HDFS 中

（2）在 spark-shell 中输入以下命令，读取 HDFS 中的数据并创建 RDD，结果如图 4-7 所示。

```
// 定义 HDFS 上"test.csv"文件的存储路径
val HDFSfile = "/Chapter4/test.csv"

// 使用 textFile 方法，加载文件并构造 RDD
val HDFSrdd = sc.textFile(HDFSfile)

// 查看返回值
HDFSrdd.collect()
```

```
scala> val HDFSfile = "/Chapter4/test.csv"
HDFSfile: String = /Chapter4/test.csv

scala> val HDFSrdd = sc.textFile(HDFSfile)
HDFSrdd: org.apache.spark.rdd.RDD[String] = /Chapter4/test.csv MapPartit
ionsRDD[8] at textFile at <console>:26

scala> HDFSrdd.collect()
res4: Array[String] = Array(Gone with the Wind,Scarlett, Pride and Preju
dice,Darcy, Rochester,Jane Eyre, The Red and the Black,Julien, The Lady
of the Camellias,Marguerite)
```

图 4-7　读取 HDFS 中的数据并创建 RDD

由图 4-7 可知，从 HDFS 的数据中创建 RDD，首先需要定义文件在 HDFS 中的存储路径，然后根据路径使用 textFile() 方法创建 RDD。

2）从本地 Linux 文件中创建 RDD

读取本地 Linux 中的文件并创建 RDD 也是通过 textFile() 方法实现的，在文件目录的前面加上"file:///"表示从本地文件系统中读取数据。

在 spark-shell 中输入以下命令，使用上述的"test.csv"文件创建 RDD，结果如图 4-8 所示。

```
val localrdd=sc.textFile("file:///root/data/Chapter4/test.csv")
localrdd.collect
```

```
scala> val localrdd=sc.textFile("file:///root/data/Chapter4/test.csv")
localrdd: org.apache.spark.rdd.RDD[String] = file:///root/data/Chapter4/
test.csv MapPartitionsRDD[10] at textFile at <console>:24

scala> localrdd.collect
res5: Array[String] = Array(Gone with the Wind,Scarlett, Pride and Preju
dice,Darcy, Rochester,Jane Eyre, The Red and the Black,Julien, The Lady
of the Camellias,Marguerite)
```

图 4-8　从本地 Linux 文件中创建 RDD

3. 创建员工业绩 RDD

（1）上传数据集到 HDFS 中。

在 master 节点中输入以下命令，上传数据文件。由于员工业绩数据集中有 3 个数据表，因此需要先将数据文件上传到本地 Linux 系统的"/root/data/Chapter4/performance/"目录下，再将整个"performance"目录上传至 HDFS 的"/Chapter4"目录下，结果如图 4-9 所示。

[root@master ~]# hdfs dfs -put /root/data/Chapter4/performance/ /Chapter4
[root@master ~]# hdfs dfs -ls /Chapter4

```
[root@master ~]# hdfs dfs -put /root/data/Chapter4/performance/ /Chapter4
2022-10-11 11:38:27,995 INFO sasl.SaslDataTransferClient: SASL encryption trus
t check: localHostTrusted = false, remoteHostTrusted = false
2022-10-11 11:38:28,199 INFO sasl.SaslDataTransferClient: SASL encryption trus
t check: localHostTrusted = false, remoteHostTrusted = false
2022-10-11 11:38:28,260 INFO sasl.SaslDataTransferClient: SASL encryption trus
t check: localHostTrusted = false, remoteHostTrusted = false
[root@master ~]# hdfs dfs -ls /Chapter4
Found 2 items
drwxr-xr-x   - root supergroup          0 2022-10-11 11:38 /Chapter4/performan
ce
-rw-r--r--   3 root supergroup        145 2022-10-11 11:24 /Chapter4/test.csv
[root@master ~]#
```

图 4-9　上传数据集到 HDFS 中

（2）从 HDFS 中读取文件并创建 RDD。

在将数据集上传到 HDFS 中之后，启动 spark-shell，并在 spark-shell 中输入以下命令，读取数据并创建员工信息 RDD 数据集 staffrdd、员工上半年业绩 RDD 数据集 firstrdd、员工下半年业绩 RDD 数据集 secondrdd，结果如图 4-10 所示。

val staffrdd=sc.textFile("/Chapter4/performance/staff.txt")
val firstrdd=sc.textFile("/Chapter4/performance/first-half-year.txt")
val secondrdd=sc.textFile("/Chapter4/performance/second-half-year.txt")

```
scala> val staffrdd=sc.textFile("/Chapter4/performance/staff.txt")
staffrdd: org.apache.spark.rdd.RDD[String] = /Chapter4/performance/staff.txt Ma
pPartitionsRDD[22] at textFile at <console>:24

scala> val firstrdd=sc.textFile("/Chapter4/performance/first-half-year.txt")
firstrdd: org.apache.spark.rdd.RDD[String] = /Chapter4/performance/first-half-y
ear.txt MapPartitionsRDD[24] at textFile at <console>:24

scala> val secondrdd=sc.textFile("/Chapter4/performance/second-half-year.txt")
secondrdd: org.apache.spark.rdd.RDD[String] = /Chapter4/performance/second-half
-year.txt MapPartitionsRDD[26] at textFile at <console>:24
```

图 4-10　从 HDFS 中读取文件并创建 RDD

4.1.2 数据操作分析

在创建 RDD 之后，就可以对 RDD 中的数据进行操作了。RDD 有两种类型的操作，一种是 Transformation 转换操作，另一种是 Actions 行动操作。

1. RDD 支持的两种操作

（1）转换操作：是指对 RDD 中的数据进行各种转换以形成新的 RDD，此操作会返回一个新的 RDD，但是它只会记录需要这样的操作，而不会立即执行，即只记录作用域 RDD 上的操作，只有在遇到一个动作（Action）时才会进行计算。常用的转换操作方法如表 4-4 所示。

RDD 的操作方法

表 4-4 常用的转换操作方法

方　法	作　用
map	将 RDD 中的每一个数据元素通过函数转换，返回新的 RDD
flatMap	首先将 map 方法应用于 RDD 中的所有元素，然后将结果扁平化拆分，返回一个新的 RDD
sortBy	通过指定条件对 RDD 中的元素进行排序
filter	通过指定条件对 RDD 中的元素进行过滤
distinct	对 RDD 中的所有元素去重，返回一个去重后的 RDD
union	将两个 RDD 合并，返回合并后的数据集
keys	返回 Pair RDD 中"键"形成的新的 RDD
values	返回 Pair RDD 中"值"形成的新的 RDD
reduceByKey	对"键"相同的值使用指定的函数进行聚合操作
groupByKey	对"键"相同的值根据指定条件进行分组
sortByKey	根据"键"对 RDD 内部的元素进行排序
join	根据"键"对两个 RDD 进行连接

（2）行动操作：是向驱动器程序返回结果或者把结果写入外部系统的操作，会触发实际的计算。常用的行动操作方法如表 4-5 所示。

表 4-5 常用的行动操作方法

方　法	作　用
count	返回 RDD 中元素的个数
take	返回 RDD 中 n 个元素的值
first	返回 RDD 中第一个元素的值
collect	返回 RDD 中所有元素的列表
top	返回 RDD 中排名前 n 的元素值

2. 统计部门人数

执行 RDD 的相关操作，对员工业绩数据集进行分析。首先，统计此部门员工的人数，可以使用 count 行动操作，返回的是 RDD 中元素的个数。

在 spark-shell 中输入以下命令，执行 count 行动操作，结果如图 4-11 所示。

```
val rdd = sc.parallelize(Seq(1,2,3,4,5,6))
rdd.count
```

```
scala> val rdd = sc.parallelize(Seq(1,2,3,4,5,6))
rdd: org.apache.spark.rdd.RDD[Int] = ParallelCollectionRDD[1] at
parallelize at <console>:24

scala> rdd.count
res1: Long = 6
```

图 4-11　count 行动操作

由图 4-11 可知，使用序列 Seq(1,2,3,4,5,6) 创建的 RDD 中有 6 个数字元素。

在创建员工业绩 RDD 的基础上进行以下操作。在 spark-shell 中输入以下命令，统计部门人数。由于部门员工的信息存储在 staffrdd 数据集中，并且每一行代表一个员工的个人信息，因此只需对其进行 count 行动操作即可，结果如图 4-12 所示。

```
staffrdd.count
```

图 4-12　统计部门人数结果

由图 4-12 可知，此部门的人数是 15 人。

3．分别统计上、下半年的业绩排名，取前 3 位和后 3 位

要获取这个问题的数据，首先需要分别对上、下半年的业绩进行排名。在创建员工业绩 RDD 时，我们使用 textFile() 方法将每行数据作为一条记录进行存储，而这里需要对业绩列进行排序，因此在排序之前需要先将数据进行转换。对业绩表中的每条数据进行分割，利用分隔符 "\t" 将每行数据分割为员工编号、时间、销售数量、业绩 4 列，并将其存储为四元组格式。销售数量、业绩列的数据类型可以通过 toInt() 方法转换为 Int 类型数据。

根据以上分析，我们可以利用 map()、flatMap()、sortBy()、take()、first()、collect() 等方法对数据进行转换操作。下面先对这些方法的使用进行解释，再实现此任务。

1）map 转换操作

map 转换操作是最常用的转换算子，可以作用在 RDD 每个分区的每个元素上，通过对 RDD 中的每个元素都执行一个指定的函数来产生一个新的 RDD，此 RDD 是分布式的数据集。

在 spark-shell 中输入以下命令，体验 map 转换操作的作用，结果如图 4-13 所示。

```
// 使用列表加载一个 RDD
```

val str1=sc.parallelize(List("I know","You know","I know that you know","I know that you know that I know"))
// 查看 str1 的内容
str1.collect
// 使用 "\t" 对每行数据进行分割
str1.map(x=>x.split("\t")).collect

```
scala> val str1=sc.parallelize(List("I know","You know","I know that you know","I know that you know that I know"))
str1: org.apache.spark.rdd.RDD[String] = ParallelCollectionRDD[11] at parallelize at <console>:24

scala> str1.collect
res6: Array[String] = Array(I know, You know, I know that you know, I know that you know that I know)

scala> str1.map(x=>x.split("\t")).collect
res7: Array[Array[String]] = Array(Array(I know), Array(You know), Array(I know that you know), Array(I know that you know that I know))
```

图 4-13 map 转换操作示例

由图 4-13 可知，在 map 转换操作中，使用了 split("\t") 函数对 str1 进行分割。

2）flatMap 转换操作

flatMap 转换操作会先将 map() 函数应用于 RDD 中的所有元素，再将返回的结果平坦化。平坦化是指返回一个元素级别全部一样的 RDD。

为了方便对比 flatMap 和 map 转换操作，对 map 转换操作中的 str1 进行 flatMap 转换操作。在 spark-shell 中输入以下命令，结果如图 4-14 所示。

str1.flatMap(x=>x.split("\t")).collect

```
scala> str1.flatMap(x=>x.split("\t")).collect
res12: Array[String] = Array(I know, You know, I know that you know, I know that you know that I know)

scala>
```

图 4-14 flatMap 转换操作示例

对比 map 和 flatMap 转换操作的输出结果，可以很容易地看出两者的区别，即 flatMap 转换操作通常用来分割单词。

3）sortBy 转换操作

sortBy 转换操作用于对 RDD 进行排序，可以输入 3 个参数，格式如下：

sortBy(参数 1, 参数 2, 参数 3)

参数 1 是要进行排序的对象值；参数 2 是排序方式，默认是正序排序，使用 false 参数就是倒序排序；参数 3 是分区个数。

在 spark-shell 中输入以下命令，体验 sortBy 转换操作，结果如图 4-15 所示。

// 加载一个 RDD，命名为 test
val test=sc.parallelize(Array(("dog",3),("cat",1),("monkey",2),("pig",3),("bird",2)))

// 根据 test 中第 2 位的元素进行降序排序

```
val test_desc = test.sortBy(_._2,false)

// 查看结果
test_desc.collect

// 根据 test 中第 2 位的元素进行升序排序
val test_asc = test.sortBy(_._2)

// 查看结果
test_asc.collect
```

```
scala> val test=sc.parallelize(Array(("dog",3),("cat",1),("monkey"
),("pig",3),("bird",2)))
test: org.apache.spark.rdd.RDD[(String, Int)] = ParallelCollection
RDD[20] at parallelize at <console>:24

scala> val test_desc = test.sortBy(_._2,false)
test_desc: org.apache.spark.rdd.RDD[(String, Int)] = MapPartitions
RDD[23] at sortBy at <console>:25

scala> test_desc.collect
res14: Array[(String, Int)] = Array((dog,3), (pig,3), (monkey,2),
(bird,2), (cat,1))

scala> val test_asc = test.sortBy(_._2)
test_asc: org.apache.spark.rdd.RDD[(String, Int)] = MapPartitionsR
DD[26] at sortBy at <console>:25

scala> test_asc.collect
res15: Array[(String, Int)] = Array((cat,1), (monkey,2), (bird,2),
 (dog,3), (pig,3))

scala>
```

图 4-15　sortBy 转换操作示例

在 sortBy 中使用了一个 "_._2" 表达式，表示对 test 对象中第 2 位的元素进行升序或降序排列，此表达式还可以写成 x=>x._2，两者意义相同，前一种是简写方式。

4）take、first、collect 行动操作

take 行动操作用于返回 RDD 中前 n 个元素的值。在 spark-shell 中输入以下命令，体验 take 行动操作。对上述 sortBy 转换操作过程中的数据集 test 进行操作，查看前两个元素的值，结果如图 4-16 所示。

test.take(2)

```
scala> test.take(2)
res18: Array[(String, Int)] = Array((dog,3), (cat,1))

scala>
```

图 4-16　take 行动操作示例

first 行动操作返回的是 RDD 中第 1 个元素的值。在 spark-shell 中输入以下命令，体验 first 行动操作。对数据集 test 进行操作，查看第 1 个元素，结果如图 4-17 所示。

test.first

```
scala> test.first
res17: (String, Int) = (dog,3)
scala>
```

图 4-17　first 行动操作示例

由以上两个输出结果可以发现，take 行动操作返回一个数组，由数据集的前 n 个元素组成；first 行动操作返回数据集中第 1 个元素的值，类似 take(1)。

collect 行动操作用于以数组的形式返回 RDD 中所有的元素。在 spark-shell 中输入以下命令，体验 collect 行动操作，对数据集 test 进行操作，结果如图 4-18 所示。

test.collect

```
scala> test.collect
res19: Array[(String, Int)] = Array((dog,3), (cat,1), (monkey,2), (pig,3), (bird,2))
scala>
```

图 4-18　collect 行动操作示例

5）任务实现

在员工业绩 RDD 的基础上进行以下操作。

（1）对上半年和下半年的业绩数据集 firstrdd、secondrdd 执行 map 转换操作，将每行数据分割为 4 列。

在 spark-shell 中输入以下命令，结果如图 4-19 所示。

val map_firstrdd=
firstrdd.map{x=>val line=x.split("\t");(line(0),line(1),line(2).toInt,line(3).toInt)}

val map_secondrdd=
secondrdd.map{x=>val line=x.split("\t");(line(0),line(1),line(2).toInt,line(3).toInt)}

```
scala> val map_firstrdd=firstrdd.map{x=>val line=x.split("\t");(line(0),line(1),line(2).toInt,line(3).toInt)}
map_firstrdd: org.apache.spark.rdd.RDD[(String, String, Int, Int)] = MapPartitionsRDD[27] at map at <console>:25

scala> val map_secondrdd=secondrdd.map{x=>val line=x.split("\t");(line(0),line(1),line(2).toInt,line(3).toInt)}
map_secondrdd: org.apache.spark.rdd.RDD[(String, String, Int, Int)] = MapPartitionsRDD[28] at map at <console>:25
```

图 4-19　使用 map 转换操作处理数据

由图 4-19 可知，使用 split("\t") 函数对每行数据进行分割，并对 line(2) 和 line(3) 使用 toInt() 方法，将字符串类型转换为 Int 数据类型。

（2）执行 sortBy 转换操作，对 map_firstrdd 和 map_secondrdd 数据集中的业绩列分别进行降序和升序排列，取出排名前 3 位和后 3 位的数据。

在 spark-shell 中输入以下命令，对上半年业绩列进行降序和升序排列，结果如图 4-20 所示。

```
val sort_firstrdd=map_firstrdd.sortBy(x=>x._4,false)
sort_firstrdd.take(3)
val sort_firstrdd=map_firstrdd.sortBy(x=>x._4)
sort_firstrdd.take(3)
```

```
scala> val sort_firstrdd=map_firstrdd.sortBy(x=>x._4,false)
sort_firstrdd: org.apache.spark.rdd.RDD[(String, String, Int, Int)
] = MapPartitionsRDD[31] at sortBy at <console>:25

scala> sort_firstrdd.take(3)
res20: Array[(String, String, Int, Int)] = Array((15012511,上半年,
20,8000), (21050309,上半年,15,6300), (21010404,上半年,18,6100))

scala> val sort_firstrdd=map_firstrdd.sortBy(x=>x._4)
sort_firstrdd: org.apache.spark.rdd.RDD[(String, String, Int, Int)
] = MapPartitionsRDD[34] at sortBy at <console>:25

scala> sort_firstrdd.take(3)
res21: Array[(String, String, Int, Int)] = Array((11012008,上半年,
4,800), (11022203,上半年,8,2300), (22010611,上半年,6,3000))

scala>
```

图 4-20　上半年业绩排名

在 spark-shell 中输入以下命令，对下半年业绩列进行降序和升序排列，结果如图 4-21 所示。

```
val sort_secondrdd=map_secondrdd.sortBy(x=>x._4,false)
sort_secondrdd.take(3)
val sort_secondrdd=map_secondrdd.sortBy(x=>x._4)
sort_secondrdd.take(3)
```

```
scala> val sort_secondrdd=map_secondrdd.sortBy(x=>x._4,false)
sort_secondrdd: org.apache.spark.rdd.RDD[(String, String, Int, Int
)] = MapPartitionsRDD[37] at sortBy at <console>:25

scala> sort_secondrdd.take(3)
res22: Array[(String, String, Int, Int)] = Array((15012511,下半年,
17,7020), (21010404,下半年,15,5800), (18090313,下半年,8,5600))

scala> val sort_secondrdd=map_secondrdd.sortBy(x=>x._4)
sort_secondrdd: org.apache.spark.rdd.RDD[(String, String, Int, Int
)] = MapPartitionsRDD[40] at sortBy at <console>:25

scala> sort_secondrdd.take(3)
res23: Array[(String, String, Int, Int)] = Array((11012008,下半年,
7,1600), (22010611,下半年,14,3000), (21110301,下半年,8,3000))

scala>
```

图 4-21　下半年业绩排名

由以上输出结果可知，上半年业绩前 3 位的是 15012511、21050309、21010404；排名后 3 位的是 11012008、11022203、22010611；下半年业绩前 3 位的是 15012511、21010404、18090313，排名后 3 位的是 11012008、22010611、21110301。

4．分别统计上、下半年业绩超过 5000 万元的员工

要获取这个问题的数据，首先需要分别过滤出上、下半年业绩超过 5000 万元的员工。

但是有些员工只有上半年或者下半年符合要求，有些员工上、下半年都符合要求，因此还需要对过滤出的员工信息进行去重，可以利用 filter()、distinct()、union() 等方法对数据进行转换操作。下面先对这些方法的使用进行解释，再实现此任务。

1）filter 转换操作

filter 转换操作用于返回满足指定过滤条件的元素，不满足条件的元素会被忽略。它需要输入一个参数，此参数是一个用于过滤的函数，该函数返回值的数据类型为 Boolean 类型，即保留返回值为 true 的元素，过滤返回值为 false 的元素。

在 spark-shell 中输入以下命令，体验 filter 转换操作，结果如图 4-22 所示。

```
// 从一个数字列表中创建 RDD
val rdd=sc.parallelize(List(1,3,6,2,8,8,5,6,7))
// 过滤大于 5 的数字
rdd.filter(x=>x>5).collect
```

```
scala> val rdd=sc.parallelize(List(1,3,6,2,8,8,5,6,7))
rdd: org.apache.spark.rdd.RDD[Int] = ParallelCollectionRDD[
13] at parallelize at <console>:24

scala> rdd.filter(x=>x>5).collect
res5: Array[Int] = Array(6, 8, 8, 6, 7)

scala>
```

图 4-22 filter 转换操作示例

2）distinct 转换操作

distinct 转换操作用于对 RDD 中的数据去重，去除完全相同的元素。在 spark-shell 中输入以下命令，体验 distinct 转换操作，对上述 filter 转换操作中的 RDD 数据进行操作，结果如图 4-23 所示。

```
rdd.distinct.collect
```

```
scala> rdd.distinct.collect
res8: Array[Int] = Array(1, 6, 3, 7, 8, 5, 2)

scala>
```

图 4-23 distinct 转换操作示例

由图 4-23 可知，原先列表中重复的数字 6 和 8 被去重了。

3）union 转换操作

union 转换操作用于将两个 RDD 进行合并，结果返回两个 RDD 的并集，并且不去重。但是要求两个 RDD 中元素的个数及元素值的数据类型保持一致。在 spark-shell 中输入以下命令，体验 union 转换操作，结果如图 4-24 所示。

```
// 创建并加载两个 RDD，分别命名为 test1 和 test2
val test1=sc.parallelize(Array(("dog",3),("cat",1),("monkey",2)))
val test2=sc.parallelize(Array(("monkey",2),("pig",3),("bird",2)))
// 执行 union 转换操作，合并两个 RDD
test1.union(test2).collect
```

```
scala> val test1=sc.parallelize(Array(("dog",3),("cat",1),("monkey",2)))
test1: org.apache.spark.rdd.RDD[(String, Int)] = ParallelCollectionRDD[31] at
 parallelize at <console>:24

scala> val test2=sc.parallelize(Array(("monkey",2),("pig",3),("bird",2)))
test2: org.apache.spark.rdd.RDD[(String, Int)] = ParallelCollectionRDD[32] at
 parallelize at <console>:24

scala> test1.union(test2).collect
res14: Array[(String, Int)] = Array((dog,3), (cat,1), (monkey,2), (monkey,2),
 (pig,3), (bird,2))
```

图 4-24 union 转换操作示例

由图 4-24 可知，创建的两个 RDD 中每个二元组的数据类型和个数均相同。通过 union 转换操作，将两个 RDD 中的所有元素合并在一起，不对重复的数据进行处理。

4）任务实现

在员工业绩 RDD 的基础上进行以下操作。

（1）按照之前讲述的方法，对 firstrdd、secondrdd 中的数据执行 map 转换操作，将每行数据分割为 4 列，结果如图 4-19 所示。

（2）执行 filter 转换操作，过滤业绩大于 5000 万元的员工，结果返回员工编号，结果如图 4-25 所示。

val filter_firstrdd=map_firstrdd.filter(x=>x._4>5000).map(x=>x._1)
val filter_secondrdd=map_secondrdd.filter(x=>x._4>5000).map(x=>x._1)

```
scala> val filter_firstrdd=map_firstrdd.filter(x=>x._4>5000).map(x=>x._1)
filter_firstrdd: org.apache.spark.rdd.RDD[String] = MapPartitionsRDD[37] at m
ap at <console>:25

scala> val filter_secondrdd=map_secondrdd.filter(x=>x._4>5000).map(x=>x._1)
filter_secondrdd: org.apache.spark.rdd.RDD[String] = MapPartitionsRDD[39] at
map at <console>:25

scala>
```

图 4-25 执行 filter 转换操作过滤业绩大于 5000 万元的员工

先通过 filter(x=>x._4>5000) 过滤业绩大于 5000 万元的员工信息，再使用 map(x=>x._1) 获取数据表中第 1 列员工编号的信息。

（3）执行 union 转换操作，对上述过滤后的数据集 filter_firstrdd 和 filter_secondrdd 进行合并，并通过 distinct 转换操作去除重复的人员编号信息。

val id=filter_firstrdd.union(filter_secondrdd).distinct()
id.collect

```
scala> val id=filter_firstrdd.union(filter_secondrdd).distinct()
id: org.apache.spark.rdd.RDD[String] = MapPartitionsRDD[47] at distinct at <con
sole>:27

scala> id.collect
res16: Array[String] = Array(19111001, 10040106, 21010404, 21050309, 15012511, 18090313)
```

图 4-26 执行 union 转换操作合并数据集

由图 4-26 输出结果可知上、下半年业绩超过 5000 万元的人员是 19111001、10040106、

21010404、21050309、15012511、18090313。

5. 统计部门当年的房屋销售套数

要获取这个问题的数据，需要先获取数据表第 3 列"房屋销售套数"的数据，再利用求和函数进行求和，可以使用 RDD 的相关描述性统计函数实现。下面先对这些函数的使用进行解释，再实现此任务。

1）min()、max() 函数

min() 函数返回 RDD 中的最小值，max() 函数返回 RDD 中的最大值。在 spark-shell 中输入以下命令，求 RDD 数据集中的最大值和最小值，结果如图 4-27 所示。

```
val rdd=sc.parallelize(List(1,3,6,2,8,8,5,6,7))
rdd.min
rdd.max
```

```
scala> val rdd=sc.parallelize(List(1,3,6,2,8,8,5,6,7))
rdd: org.apache.spark.rdd.RDD[Int] = ParallelCollectionR
DD[54] at parallelize at <console>:24

scala> rdd.min
res18: Int = 1

scala> rdd.max
res19: Int = 8
```

图 4-27　min()、max() 函数操作结果

2）mean()、sum() 函数

mean() 函数返回 RDD 中的平均值。sum() 函数返回 RDD 中的总和。在 spark-shell 中输入以下命令，求 RDD 数据集的平均值和总和，结果如图 4-28 所示。

```
rdd.mean()
rdd.sum()
```

```
scala> rdd.mean()
res20: Double = 5.111111111111111

scala> rdd.sum()
res21: Double = 46.0

scala>
```

图 4-28　mean()、sum() 函数操作结果

3）variance()、stdev() 函数

variance() 函数计算 RDD 中所有元素的总体方差。stdev() 函数计算 RDD 中所有元素的样本方差。在 spark-shell 中输入以下命令，求 RDD 数据集的总体方差和样本方差，结果如图 4-29 所示。

```
rdd.variance()
rdd.stdev()
```

```
scala> rdd.variance()
res24: Double = 5.876543209876543

scala> rdd.stdev()
res25: Double = 2.4241582476968255

scala>
```

图 4-29　variance()、stdev() 函数操作结果

总体方差和样本方差是统计学中的两个概念，在数据处理和分析中具有重要作用。

4）任务实现

在员工业绩 RDD 的基础上进行以下操作。

（1）按照上述方法，对数据进行 map 转换操作，将每行数据分割为 4 列，结果如图 4-19 所示。

（2）将经过 map 转换操作处理的上、下半年业绩数据集 map_firstrdd 和 map_secondrdd 进行合并，取出第 3 列数据，利用 sum() 函数计算房屋销售总套数。

在 spark-shell 中输入以下命令，结果如图 4-30 所示。

val housetotal=map_firstrdd.union(map_secondrdd).map(x=>x._3).sum

```
scala> val housetotal=map_firstrdd.union(map_secondrdd).map(x=>x._3).sum
housetotal: Double = 352.0
```

图 4-30　房屋销售总套数

由图 4-30 可知，此部门当年的房屋销售总套数为 352 套。

6. 查看全年总业绩最高的员工

此任务需要将每个员工上、下半年的业绩相加并取最高者。这个过程需要对员工编号相同的业绩数据进行识别和相加。在 RDD 中可以通过键－值对的操作完成。

Spark 为包含键－值对的 RDD 提供了一些专有的操作，这些 RDD 被称为 Pair RDD。接下来介绍有关键－值对 RDD 的操作。

1）创建 Pair RDD

创建 Pair RDD 的方法有很多，当需要将一个普通 RDD 转换为 Pair RDD 时，可以使用 map 转换操作来实现。

在 spark-shell 中输入以下命令，创建一个 Pair RDD，结果如图 4-31 所示。

Pair RDD 的创建和操作方法

```
// 创建一个 RDD
val rdd = sc.parallelize(List("dog","cat","monkey","pig"))
// 创建键－值对
val pairrdd = rdd.map(word => (word,1))
// 查看结果
pairrdd.collect
```

```
scala> val rdd = sc.parallelize(List("dog","cat","monkey","pig"))
rdd: org.apache.spark.rdd.RDD[String] = ParallelCollectionRDD[70] at parallelize
 at <console>:24

scala> val pairrdd = rdd.map(word => (word,1))
pairrdd: org.apache.spark.rdd.RDD[(String, Int)] = MapPartitionsRDD[71] at map a
t <console>:25

scala> pairrdd.collect
res28: Array[(String, Int)] = Array((dog,1), (cat,1), (monkey,1), (pig,1))
```

图 4-31　创建 Pair RDD

由图 4-31 可知，使用 map(word => (word,1)) 将 RDD 中的每个元素都作为一个键，并赋予一个值 "1"，结果会返回一个 Pair RDD。

2）keys、values 转换操作

keys 转换操作返回的类型是 RDD，内部元素为每个键-值对中的键。values 转换操作返回的类型也是 RDD，内部元素为每个键-值对中的值。在 spark-shell 中输入以下命令，对上一个任务的键-值对 RDD 执行 keys、values 转换操作，结果如图 4-32 所示。

```
// 取 Pair RDD 中的键
pairrdd.keys.collect
// 取 Pair RDD 中的值
pairrdd.values.collect
```

```
scala> pairrdd.keys.collect
res32: Array[String] = Array(dog, cat, monkey, pig)

scala> pairrdd.values.collect
res33: Array[Int] = Array(1, 1, 1, 1)
```

图 4-32　keys、values 转换操作结果

3）reduceByKey 转换操作

reduceByKey 转换操作应用于键-值对数据集中，对键相同的值使用指定的函数进行聚合操作，返回一个键-值对的数据集。在 spark-shell 中输入以下命令，体验 reduceByKey 转换操作，结果如图 4-33 所示。

```
// 创建普通 RDD
val test=sc.parallelize(Array(("dog",3),("cat",1),("pig",2),("pig",3),("dog",2)))

// 创建 Pair RDD
val map_test=test.map(x=>(x._1,x._2))

// 使用 reduceByKey 对键相同的值进行相加操作
val reduce_test=map_test.reduceByKey((x,y)=>x+y)
// 查看结果
reduce_test.collect
```

由图 4-33 可知，使用 reduceByKey((x,y)=>x+y) 对 Pair RDD 中键相同的值进行相加操作。在此例中，相同的键有 "dog" 和 "pig"，对其值进行相加操作，最终获得 Array((dog,5),(pig,5),(cat,1)) 的结果。

第 4 章 房产大数据分析与探索

```
scala> val test=sc.parallelize(Array(("dog",3),("cat",1),("pig",2),("pig",3),("d
og",2)))
test: org.apache.spark.rdd.RDD[(String, Int)] = ParallelCollectionRDD[76] at par
allelize at <console>:24

scala> val map_test=test.map(x=>(x._1,x._2))
map_test: org.apache.spark.rdd.RDD[(String, Int)] = MapPartitionsRDD[77] at map
at <console>:25

scala> val reduce_test=map_test.reduceByKey((x,y)=>x+y)
reduce_test: org.apache.spark.rdd.RDD[(String, Int)] = ShuffledRDD[78] at reduce
ByKey at <console>:25

scala> reduce_test.collect
res34: Array[(String, Int)] = Array((dog,5), (pig,5), (cat,1))
```

图 4-33　reduceByKey 转换操作结果

4）groupByKey 转换操作

groupByKey 转换操作会对具备相同键的值进行分组，形成二元元组，第 1 个字段为相同的键，第 2 个字段为具备相同键的值的集合。使用此操作通常是为了在分组之后对具备相同键的值进行统计等操作。

在 spark-shell 中输入以下命令，体验 groupByKey 转换操作，对上一个任务中的 map_test 数据集进行操作，结果如图 4-34 所示。

val group_test=map_test.groupByKey()
group_test.collect

```
scala> val group_test=map_test.groupByKey()
group_test: org.apache.spark.rdd.RDD[(String, Iterable[Int])] = ShuffledRDD[2] a
t groupByKey at <console>:25

scala> group_test.collect
res0: Array[(String, Iterable[Int])] = Array((dog,CompactBuffer(3, 2)), (pig,Com
pactBuffer(2, 3)), (cat,CompactBuffer(1)))
```

图 4-34　groupByKey 转换操作结果

通过上一个操作可知，在 map_test 数据集中，有相同键的元组是 ("dog",3) 和 ("dog",2)、("pig",2) 和 ("pig",3)。由图 4-34 可知，在执行 groupByKey 转换操作之后，分别形成了 (dog,CompactBuffer(3,2)) 和 (pig,CompactBuffer(2,3))。虽然 ("cat",1) 没有相同键的元组，但是其值也会转换为一个集合。

5）sortByKey 转换操作

sortByKey 转换操作返回一个根据键进行排序的 RDD。在 spark-shell 中输入以下命令，体验 sortByKey 转换操作，结果如图 4-35 所示。

val pairRdd = sc.parallelize(List(("c",3),("b",2),("a",1)))
pairRdd.sortByKey().collect

```
scala> val pairRdd = sc.parallelize(List(("c",3),("b",2),("a",1)))
pairRdd: org.apache.spark.rdd.RDD[(String, Int)] = ParallelCollectionRDD[5] at p
arallelize at <console>:24

scala> pairRdd.sortByKey().collect
res2: Array[(String, Int)] = Array((a,1), (b,2), (c,3))
```

图 4-35　sortByKey 转换操作结果

由图 4-35 可知，sortByKey 转换操作会根据键对应的值进行排序，并且默认是升序排序，若想对值进行降序排序，则需要在后面加上 false 参数，即 sortByKey(false)。

6）任务实现

在员工业绩 RDD 的基础上进行以下操作。

（1）将上、下半年的业绩合并到同一个 RDD 中。

按照之前讲述的方法，对数据进行 map 转换操作，将每行数据分割为 4 列，结果如图 4-19 所示。

对 map_firstrdd 和 map_secondrdd 数据集进行合并操作，结果如图 4-36 所示。

val total=map_firstrdd.union(map_secondrdd)

```
scala> val total=map_firstrdd.union(map_secondrdd)
total: org.apache.spark.rdd.RDD[(String, String, Int, Int)] = UnionRDD[16] at un
ion at <console>:27
```

图 4-36　使用 union 转换操作合并两个数据集

（2）将数据转换成 (员工编号 , 业绩) 键 - 值对，通过 reduceByKey 转换操作对相同员工编号对应的业绩值进行相加操作，结果如图 4-37 所示。

val yeji=total.map(x=>(x._1,x._4)).reduceByKey((x,y)=>x+y)
yeji.collect

```
scala> val yeji=total.map(x=>(x._1,x._4)).reduceByKey((x,y)=>x+y)
yeji: org.apache.spark.rdd.RDD[(String, Int)] = ShuffledRDD[18] at reduceByKey a
t <console>:25

scala> yeji.collect
res3: Array[(String, Int)] = Array((19111001,11410), (19080303,8600), (10040106,
9020), (20100706,9000), (11040204,8000), (21010404,11900), (21050309,11600), (12
080403,8300), (15012511,15020), (11020101,7101), (11012008,2400), (22010611,6000
), (11022203,5600), (21110301,6000), (18090313,11600))
```

图 4-37　计算业绩总和

由图 4-37 可知，先通过 map 转换操作获取 total 数据集中的第 1 列员工编号和第 4 列业绩，形成 Pair RDD；再通过 reduceByKey 转换操作处理并获得各员工的年度业绩总和。输出的结果依然是键 - 值对 RDD，包含员工编号和业绩总和。

（3）对业绩总和进行排序，获得最高业绩的员工信息，结果如图 4-38 所示。

// 通过 sortBy 转换操作对指定列进行排序
yeji.sortBy(x=>x._2,false)
yeji.sortBy(x=>x._2,false).take(1)

```
scala> yeji.sortBy(x=>x._2,false)
res7: org.apache.spark.rdd.RDD[(String, Int)] = MapPartitionsRDD[23] at sortBy a
t <console>:26

scala> yeji.sortBy(x=>x._2,false).take(1)
res8: Array[(String, Int)] = Array((15012511,15020))
```

图 4-38　最高业绩员工信息

7. 存储以上统计分析信息

数据存储的方式有多种，比如将数据存储为 JSON 文件、CSV 文件、文本文件等。本任务将对上、下半年的业绩及全年总业绩数据集进行 join 连接，并将结果数据以文本文件的形式存储到 HDFS 中。

数据 join 连接的操作主要有 join、rightOuterJoin、leftOuterJoin、fullOuterJoin，其操作与 SQL 中的 join 操作含义一致。

文本文件的存储可以直接调用 saveAsTextFile(path)。

在员工业绩 RDD 的基础上进行以下操作。

（1）对数据执行 map 转换操作，对每行数据以"\t"进行分割，结果如图 4-39 所示。

```
// 对 map_staf 执行 map 转换操作，分割为两列，分别是员工编号和姓名
val map_staff=staffrdd.map{x=>val line=x.split("\t");(line(0),line(1))}
// 对 map_firstrdd 执行 map 转换操作，获取员工编号和业绩列
val map_firstrdd=firstrdd.map{x=>val line=x.split("\t");(line(0),line(3).toInt)}
// 对 map_secondrdd 执行 map 转换操作，获取员工编号和业绩列
val map_secondrdd=secondrdd.map{x=>val line=x.split("\t");(line(0),line(3).toInt)}
```

```
scala> val map_staff=staffrdd.map{x=>val line=x.split("\t");(line(0),line(1))}
map_staff: org.apache.spark.rdd.RDD[(String, String)] = MapPartitionsRDD[56] at map at <console>:25

scala> val map_firstrdd=firstrdd.map{x=>val line=x.split("\t");(line(0),line(3).toInt)}
map_firstrdd: org.apache.spark.rdd.RDD[(String, Int)] = MapPartitionsRDD[57] at map at <console>:25

scala> val map_secondrdd=secondrdd.map{x=>val line=x.split("\t");(line(0),line(3).toInt)}
map_secondrdd: org.apache.spark.rdd.RDD[(String, Int)] = MapPartitionsRDD[58] at map at <console>:25
```

图 4-39　对数据执行 map 转换操作结果

（2）使用 join 连接，将员工信息及上、下半年的业绩数据集进行连接，结果如图 4-40 所示。

```
val total1=map_staff.join(map_firstrdd).join(map_secondrdd)
```

```
scala> val total1=map_staff.join(map_firstrdd).join(map_secondrdd)
total1: org.apache.spark.rdd.RDD[(String, ((String, Int), Int))] = MapPartitionsRDD[64] at join at <console>:29
```

图 4-40　连接员工信息及业绩数据集结果

（3）继续使用 join 连接添加年度总业绩，结果如图 4-41 所示。

```
val total2=total1.join(yeji)
total2.collect
```

```
scala> val total2=total1.join(yeji)
total2: org.apache.spark.rdd.RDD[(String, (((String, Int), Int), Int))] = MapPar
titionsRDD[67] at join at <console>:27

scala> total2.collect
res13: Array[(String, (((String, Int), Int), Int))] = Array((19111001,(((尚梦菲,
5900),5510),11410)), (19080303,(((任晓燕,4300),4300),8600)), (10040106,(((王向秋
,5010),4010),9020)), (20100706,(((杨昆明,4800),4200),9000)), (11040204,(((黄焱,3
500),4500),8000)), (21010404,(((李安安,6100),5800),11900)), (21050309,(((李悦可,
6300),5300),11600)), (12080403,(((沈睿广,4200),4100),8300)), (15012511,(((王嘉勋
,8000),7020),15020)), (11020101,(((马文轩,3400),3701),7101)), (11012008,(((徐悠
,800),1600),2400)), (22010611,(((许辉,3000),3000),6000)), (11022203,(((刘聪,2300)
,3300),5600)), (21110301,(((刘浩丽,3000),3000),6000)), (18090313,(((廖文敏,6000)
,5600),11600)))
```

图 4-41　join 连接添加年度总业绩

由以上输出结果可知，在将员工信息，上、下半年的业绩数据集，年度总业绩进行 join 连接之后，形成的数组中的元素处于不同级别，不利于后续的操作与存储，因此需要对 total2 执行 map 转换操作。

（4）对上述结果 total2 中的数据执行 map 转换操作，使所有元素都成为同一级别的元素，结果如图 4-42 所示。

val total=total2.map(x=>Array(x._1,x._2._1._1._1,x._2._1._1._2,x._2._1._2,x._2._2).mkString(","))
total.collect

```
scala> val total=total2.map(x=>Array(x._1,x._2._1._1._1,x._2._1._1._2,x._2._1._2
,x._2._2).mkString(","))
total: org.apache.spark.rdd.RDD[String] = MapPartitionsRDD[68] at map at <consol
e>:25

scala> total.collect
res14: Array[String] = Array(19111001,尚梦菲,5900,5510,11410, 19080303,任晓燕,43
00,4300,8600, 10040106,王向秋,5010,4010,9020, 20100706,杨昆明,4800,4200,9000, 11
040204,黄焱,3500,4500,8000, 21010404,李安安,6100,5800,11900, 21050309,李悦可,630
0,5300,11600, 12080403,沈睿广,4200,4100,8300, 15012511,王嘉勋,8000,7020,15020, 1
1020101,马文轩,3400,3701,7101, 11012008,徐悠,800,1600,2400, 22010611,许辉,3000,3
000,6000, 11022203,刘聪,2300,3300,5600, 21110301,刘浩丽,3000,3000,6000, 18090313
,廖文敏,6000,5600,11600)
```

图 4-42　执行 map 转换操作结果

（5）将 total 数据集中的数据存储为文本文件，并保存到 HDFS 中。通过 repartition() 方法重新设置分区为 1。

在 spark-shell 中输入以下命令：

total.repartition(1).saveAsTextFile("/Chapter4/total")

输出结果为空行，则表示存储成功，如图 4-43 所示。

```
scala> total.repartition(1).saveAsTextFile("/Chapter4/total")

scala>
```

图 4-43　将数据存储为文本文件并保存到 HDFS 中

（6）在 HDFS 中查看保存的结果，在 master 节点中输入以下命令，结果如图 4-44 所示。

[root@master ~]# hdfs dfs -cat /Chapter4/total/part-00000

第 4 章 房产大数据分析与探索

```
[root@master ~]# hdfs dfs -cat /Chapter4/total/part-00000
2022-10-13 07:37:09,769 INFO sasl.SaslDataTransferClient: SASL e
ncryption trust check: localHostTrusted = false, remoteHostTrust
ed = false
19111001,尚梦菲,5900,5510,11410
19080303,任晓燕,4300,4300,8600
10040106,王向秋,5010,4010,9020
20100706,杨昆明,4800,4200,9000
11040204,黄焱,3500,4500,8000
21010404,李安安,6100,5800,11900
21050309,李悦可,6300,5300,11600
12080403,沈睿广,4200,4100,8300
15012511,王嘉勋,8000,7020,15020
11020101,马文轩,3400,3701,7101
11012008,徐悠,800,1600,2400
22010611,许辉,3000,3000,6000
11022203,刘聪,2300,3300,5600
21110301,刘浩丽,3000,3000,6000
18090313,廖文敏,6000,5600,11600
```

图 4-44 在 HDFS 中查看保存的数据结果

由图 4-44 可知，汇总的数据分别是员工编号，姓名，上、下半年的业绩及年度总业绩。

任务 4.2 某城市近年房产销售状况分析

案例分析——某城市近年房产销售状况分析

情境导入

房子是一种昂贵的"商品"，但在每个人生活中都是必不可少的。房子的价格、所处的位置、装修情况等因素都是购房者关心的问题。了解房产市场的变化规律以及各个因素对价格的影响，对于购房者来说是购房之前首先要做的功课。大数据专业的学生应该思考如何用所学知识服务社会，在增进个人技能的同时，解决生活中的热点问题。

现有一份来自某网站的某城市近年房产销售数据集（见"house-price.csv"文件），包含交易时间、总价、装修情况、所属区域等信息。具体字段及其含义如表 4-6 所示。

表 4-6 某城市近年房产销售数据集字段及其含义

字 段	含 义
url	数据链接
id	用户 ID
tradeTime	交易时间
totalPrice	总价（万元）
price	单价
square	面积
renovationCondition	装修情况

续表

字 段	含 义
elevator	是否有电梯
fiveYearsProperty	是否满五年
subway	是否有地铁
district	所属区域
communityAverage	区域均价

通过数据分析解决以下问题：

1．近几年房产销售量趋势如何？

2．此城市各区域房产销售量如何？均价如何？

3．在 2018 年 1 月 1 日的销售量有多少？

4．查询 2018 年 1 月 1 日到 1 月 31 日之间，满五年房产的销售量。

5．哪种装修类型的房子销售量最高？

6．查询所售卖的房子中，电梯有无的比例、地铁有无的情况。

学习目标和要求

知识与技能目标

能灵活、综合地应用各种 RDD 操作和各个算子对数据进行分析。

素质目标

具有用所学知识服务社会的责任意识与奉献精神。

4.2.1 数据准备

此房产销售数据集是一个在系统外部的 CSV 文件，因此需要先将其上传至文件系统中，加载为 RDD 后再做处理分析。

（1）利用 MobaXterm 工具，将 "house-price.csv" 文件上传到本地 Linux 的 "/root/data/Chapter4" 目录下，结果如图 4-45 所示。

图 4-45　上传数据文件到本地

（2）启动 Hadoop 集群和 Spark 集群，将 "/root/data/Chapter4" 目录下的 "house-price.csv" 文件上传到 HDFS 的 "/Chapter4" 目录下。

第 4 章 房产大数据分析与探索

在 master 节点中输入以下命令,将文件上传到 HDFS 中,结果如图 4-46 所示。

[root@master ~]# hdfs dfs -mkdir /Chapter4　# 若此目录已存在,则无须操作
[root@master ~]# hdfs dfs -put /root/data/Chapter4/house-price.csv /Chapter4

```
[root@master ~]# hdfs dfs -mkdir /Chapter4
[root@master ~]# hdfs dfs -put /root/data/Chapter4/house-price.csv /Chapter4
2022-10-14 01:07:35,572 INFO sasl.SaslDataTransferClient: SASL encryption trust check: localHost
Trusted = false, remoteHostTrusted = false
```

图 4-46　上传文件到 HDFS 中

在 master 节点中输入以下命令,查看文件是否上传成功,结果如图 4-47 所示,表示上传成功。

[root@master ~]# hdfs dfs -ls /Chapter4

```
[root@master ~]# hdfs dfs -ls /Chapter4
Found 1 items
-rw-r--r--   3 root supergroup   42149699 2022-10-14 01:07 /Chapter4/house-price.csv
[root@master ~]#
```

图 4-47　查看文件是否上传成功

(3) 启动 spark-shell,加载数据到 RDD 中,查看 1 行数据,并了解数据内容。

首先,定义一个放置数据集文件的 HDFS 路径(filepath);然后,利用 sc.textFile (filepath) 创建名称为 housepriceRDD 的 RDD;最后,利用 take(1) 查看 1 行数据。

在 spark-shell 中输入以下命令,结果如图 4-48 所示。

```
// 定义放置数据集文件的 HDFS 路径
val filepath = "/Chapter4/house-price.csv"
// 将数据集读取到 RDD 中
val housepriceRDD = sc.textFile(filepath)
// 查看 1 行数据
housepriceRDD.take(1)
```

```
scaval filepath = "/Chapter4/house-price.csv"
filepath: String = /Chapter4/house-price.csv

scala> val housepriceRDD = sc.textFile(filepath)
housepriceRDD: org.apache.spark.rdd.RDD[String] = /Chapter4/house-price.csv MapPartit
ionsRDD[9] at textFile at <console>:26

scala>  housepriceRDD.take(1)
res3: Array[String] = Array(https://bj.lianjia.com/chengjiao/101084782030.html,BJ1000
011001,2016/8/9,415,31680,131,简装房,有电梯,未满五年,有地铁,朝阳区,56021)
```

图 4-48　查看 1 行数据

由图 4-48 可知,此数组中第 0 个元素是此行数据的链接,第 1 个元素是用户 ID,第 2 个元素是交易时间,第 3 个元素是总价,第 4 个元素是单价,第 5 个元素是面积,第 6 个元素是装修情况,第 7 个元素是是否有电梯,第 8 个元素是是否满五年,第 9 个元素是是否有地铁,第 10 个元素是所属区域,第 11 个元素是区域均价。

133

4.2.2 数据探索与分析

在进行数据探索分析之前，需要先将加载到 RDD 中的数据进行转换，因为在 housepriceRDD 中，每一条数据都属于一个元素整体，需要对每一条数据根据字段含义进行分割，以便后续的数据处理和分析；再利用弹性分布式数据集的各个转换及行动操作进行分析统计。

1. 对 RDD 进行转换操作

使用 map 转换操作转换 RDD，以","将每一行数据分割，并查看数据情况。

在 spark-shell 中输入以下命令，查看数据集中的两条数据，结果如图 4-49 所示。

```
val m_houseprice = housepriceRDD.map(line => line.split(","))
m_houseprice.take(2)
```

图 4-49　查看数据情况

2. 统计用户人数

对于房产企业的销售人员，需要对用户的情况有基本的了解。下面对此份数据集中用户人数的基本情况进行统计，获得记录条目的总数，以及在这些记录中包括多少个用户 ID。

可以利用 count 行动操作统计 m_houseprice 数据集中的记录总数，利用 distinct 转换操作对 id 列进行去重和数目统计即可获得用户 ID 数量。

在 spark-shell 中输入以下命令，结果如图 4-50 所示。

```
// 查看数据总条数
m_houseprice.count
// 查看这些记录中的用户 ID 数量
m_houseprice.map(_(1)).distinct.count
```

图 4-50　用户 ID 数量结果

由图 4-50 可知，此数据集共有 318851 条数据，来自 226636 位用户。

3. 统计近几年房产销售量情况

要获得近几年的房产销售量情况、了解房产销售量趋势，需要先从日期中提取年份数据，再利用 reduceByKey 转换操作按年进行统计汇总，最后利用 sortBy 转换操作对统计汇总的数据进行排序，获得近几年销售量从高到低的排序情况。

在 spark-shell 中使用 :paste 模式输入以下命令，结果如图 4-51 所示。

```
// 导入相关依赖包
import java.text.SimpleDateFormat
import java.util.Date
import java.util.Calendar

// 定义一个 tranfTime() 函数，用来提取日期中的年份
def tranfTime(timestring: String): Int = {
  val fm = new SimpleDateFormat("yyyy/MM/dd")
  // 将 string 的时间转换为 date
  val time: Date = fm.parse(timestring)
  val cal = Calendar.getInstance()
  cal.setTime(time)
  // 提取时间中的年份
  val year: Int = cal.get(Calendar.YEAR)
  year
}

// 对 m_houseprice 进行分析
m_houseprice.map(arr => (tranfTime(arr(2)),1)).     // 生成 ( 年份 ,1) 元组
    reduceByKey(_ + _).                              // 按年进行统计汇总
    sortBy(_._2,false).                              // 按销售量进行排序
    take(9).foreach(println)                         // 查看前 9 条记录
```

图 4-51　近几年房产销售量情况

由图 4-51 可知，在 2010 年到 2018 年之间，2016 年的房产销售量最高。2016 年之前呈现上涨趋势，之后呈现下降趋势。

4. 分析房屋所属区域、各区域的房产销售量、区域均价

1）房屋所属区域及销售量情况

在 m_houseprice 数据集中，"所属区域"数据在第 10 列，在使用 distinct 转换操作对该列去重之后就能获得购买房屋所属的各区域的情况。对于各区域的销售量，可以通过使用 reduceByKey 转换操作，对具有相同键（区域）的值进行相加操作来获得。

在 spark-shell 中输入以下命令，结果如图 4-52 所示。

```
// 查看区域数量
m_houseprice.map(_(10)).distinct.count
// 查看有哪些区域
m_houseprice.map(_(10)).distinct.collect
// 统计各区域的房产销售量
m_houseprice.map(arr => (arr(10),1)).reduceByKey(_+_).collect.foreach(println)
```

```
scala> m_houseprice.map(_(10)).distinct.count
res21: Long = 13

scala> m_houseprice.map(_(10)).distinct.collect
res22: Array[String] = Array(平谷区, 丰台区, 东城区, 朝阳区, 门头沟区, 石景山区, 房山区, 西城区, 昌平区, 大兴区, 通州区, 顺义区, 海淀区)

scala> m_houseprice.map(arr => (arr(10),1)).reduceByKey(_+_).collect.foreach(println)
(平谷区,13974)
(丰台区,29338)
(东城区,17086)
(朝阳区,107244)
(门头沟区,1704)
(石景山区,11371)
(房山区,2955)
(西城区,31293)
(昌平区,38634)
(大兴区,15313)
(通州区,2537)
(顺义区,9202)
(海淀区,38200)
```

图 4-52　各区域的房产销售量情况

由图 4-52 可知，此数据集中所售卖的房产来自 13 个区域，各区域的销售量情况也清晰明了。

2）各区域的房屋均价情况

除了关注房屋销售量，各区域房屋均价情况也是购房者关注的重点。

（1）使用 map 转换操作取出"所属区域""区域均价"两列数据，此处的"区域均价"是指各个时间点的房屋均价，对各时间点的均价求平均值即可得到此区域的房屋均价。将"区域均价"转换为 Double 类型的数据，并对每个价格计数"1"，目的是便于之后统计 Key 的个数。在 spark-shell 中输入以下命令，结果如图 4-53 所示。

```
val price1=m_houseprice.map(arr => (arr(10),(arr(11).toDouble,1)))
// 查看处理结果
price1.take(5).foreach(println)
```

```
scala> val price1=m_houseprice.map(arr => (arr(10),(arr(11
).toDouble,1)))
price1: org.apache.spark.rdd.RDD[(String, (Double, Int))]
= MapPartitionsRDD[112] at map at <console>:25

scala> price1.take(5).foreach(println)
(朝阳区,(56021.0,1))
(朝阳区,(71539.0,1))
(朝阳区,(48160.0,1))
(昌平区,(51238.0,1))
(东城区,(62588.0,1))
```

图 4-53　获取区域的房屋均价

（2）对相同区域的均价进行相加操作。在 spark-shell 中输入以下命令，结果如图 4-54 所示。

```
//x._1+y._1 是指将"区域"键相同的"均价"值相加，x._2+y._2 是统计相同区域的数量
val price2=price1.reduceByKey((x,y)=>(x._1+y._1,x._2+y._2))
// 查看操作结果
price2.take(5).foreach(println)
```

```
scala> val price2=price1.reduceByKey((x,y)=>(x._1+y._1,x._
2+y._2))
price2: org.apache.spark.rdd.RDD[(String, (Double, Int))]
= ShuffledRDD[113] at reduceByKey at <console>:25

scala> price2.take(5).foreach(println)
[Stage 76:>
(平谷区,(6.14591873E8,13974))
(丰台区,(1.613471401E9,29338))
(东城区,(1.53261012E9,17086))
(朝阳区,(6.746697601E9,107244))
(门头沟区,(6.6141006E7,1704))
```

图 4-54　相同区域的均价相加操作

（3）求平均房价，并排序输出。

在 spark-shell 中输入以下命令，结果如图 4-55 所示。

```
// 将"均价"值总和除以相同"区域"键的数量即可获得均价
val price3=price2.map(a=>(a._1,a._2._1/a._2._2))
// 对均价进行降序排序
val price4=price3.sortBy(x=>x._2,false).collect.foreach(println)
```

```
scala> val price3=price2.map(a=>(a._1,a._2._1/a._2._2))
price3: org.apache.spark.rdd.RDD[(String, Double)] = MapPa
rtitionsRDD[114] at map at <console>:25

scala> val price4=price3.sortBy(x=>x._2,false).collect.for
each(println)
(西城区,101701.7312817563)
(东城区,89699.76120800656)
(海淀区,79649.65562827224)
(朝阳区,62909.79076684943)
(丰台区,54995.95749539846)
(石景山区,50594.79482895084)
(通州区,47452.02404414663)
(大兴区,44417.67818193691)
(平谷区,43981.09868326893)
(昌平区,43030.56838536004)
(顺义区,39402.48217778744)
(门头沟区,38815.144366197186)
(房山区,35706.79627749577)
price4: Unit = ()
```

图 4-55　各个区域的房屋均价排名

由图 4-55 可知各个区域的房屋均价，最高房屋均价与最低房屋均价间甚至有近 3 倍的差距，说明地段对于房价的影响还是十分明显的。

5. 查询 2018 年 1 月 1 日的日销量

查询某一天或某一条件下的数据，可以使用 filter 转换操作。在 m_houseprice 数据集中，"交易时间"数据在第 2 列，在 spark-shell 中输入以下命令，结果如图 4-56 所示。

m_houseprice.filter(_(2)=="2018/1/1").count

```
scala> m_houseprice.filter(_(2)=="2018/1/1").count
res25: Long = 46
```

图 4-56　某一天的销售情况

由图 4-56 可知，2018 年 1 月 1 日这一天的房屋日销售量是 46 套。

6. 查询 2018 年 1 月 1 日到 1 月 31 日之间，满五年房产的销售量

使用 filter 转换操作，可以圈定某一条件范围，在 spark-shell 中输入以下命令，结果如图 4-57 所示。

m_houseprice.filter(_(8)==" 满五年 ").
　　filter(_(2)>="2018/1/1").
　　filter(_(2)<"2018/1/31").
　　count

```
scala> m_houseprice.filter(_(8)=="满五年").
     |           filter(_(2)>="2018/1/1").
     |           filter(_(2)<"2018/1/31").
     |           count
res27: Long = 131
```

图 4-57　某个月的销售情况

由图 4-57 可知，2018 年 1 月 1 日到 1 月 31 日之间，满五年房产的销售数量是 131 套。

7. 分析哪种装修类型的房子销售量高

在数据集中，房子的装修有精装、简装、毛坯、其他这 4 种类型。此问题的分析思路与统计区域房产销售量情况一致，可以使用 reduceByKey(_+_) 对装修类型相同的房屋进行求和统计。在 m_houseprice 数据集中，"装修类型"数据在第 6 列。在 spark-shell 中输入以下命令，结果如图 4-58 所示。

m_houseprice.map(arr => (arr(6),1)).reduceByKey(_+_).collect.foreach(println)

```
scala> m_houseprice.map(arr => (arr(6),1)).reduceByKey(_+_).collect.foreach(println)
(毛坯房,5390)
(精装房,117438)
(其他,118772)
(简装房,77251)
```

图 4-58　不同装修类型的房子销售量情况

由图 4-58 可知，相较于毛坯房和简装房，精装房的销售量高很多。

第 4 章　房产大数据分析与探索

8. 分析是否有电梯、是否有地铁情况

在 m_houseprice 数据集中,"是否有电梯"数据在第 7 列,"是否有地铁"数据在第 9 列。首先,使用 map(arr => (arr(7),1)) 对电梯数据列中的每个元素创建键-值对,其中"键"指有无电梯,为每个"值"赋予"1"。然后,使用 groupByKey 转换操作对相同"键"的键-值对进行分类,并对"值"进行求和。地铁数据列的处理思路相同。

在 spark-shell 中输入以下命令,结果如图 4-59 所示。

```
m_houseprice.map(arr => (arr(7),1)).groupByKey.map(x=>(x._1,x._2.sum)).collect
m_houseprice.map(arr => (arr(9),1)).groupByKey.map(x=>(x._1,x._2.sum)).collect
```

```
scala> m_houseprice.map(arr => (arr(7),1)).groupByKey.map(x=>(x._1,x._2.sum)).collect
res5: Array[(String, Int)] = Array((无电梯,134875), (有电梯,183976))

scala> m_houseprice.map(arr => (arr(9),1)).groupByKey.map(x=>(x._1,x._2.sum)).collect
res6: Array[(String, Int)] = Array((无地铁,127205), (有地铁,191646))
```

图 4-59　是否有电梯和地铁情况

4.2.3　总结分析

根据对"house-price.csv"中数据集的分析,可知此城市房价呈现上涨趋势,不同区域房价差异明显,最高均价区域与最低均价区域间甚至有近 3 倍的差距,说明地段对于房价的影响还是十分明显的。购房者在买房时首要考虑的还是交通问题,房屋在地铁沿线上的售出率高;其次,房屋电梯情况也是要考虑的问题,有电梯的房屋销售量高于无电梯的房屋。此外,房屋的装修情况,除了"其他"类型的房子,精装房更受购房者欢迎。

脑图小结

本章介绍了 Spark 中弹性分布式数据集 RDD 的创建方法、利用 RDD 的各个算子对数据进行操作分析的方法。通过"某房产公司销售人员业绩分析"和"某城市近年房产销售状况分析"两个任务,灵活、综合地应用各种 RDD 操作和各个算子对数据进行分析处理。通过以下脑图小结,助力学习者掌握和巩固相关知识。

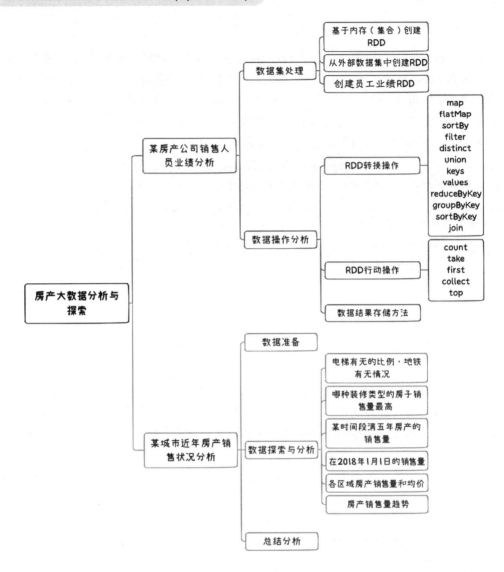

章节练习

现有一份某售房网站的房屋销售情况,包括用户编号(id)、交易时间(tradeTime)、房屋总价(totalPrice)、房屋单价(price)、房屋面积(square)、卧室数量(livingRoom)、楼层情况(level)、具体楼层(floor)、建筑时间(constructionTime)、梯户比(ladderRatio)。根据本章知识,对以下问题进行分析:

1. 查看数据集中售房楼层的高、中、低、底、顶这几种类型,分析各个楼层的房屋销量情况;

2. 分析在所售卖的房屋中,卧室是几室的房屋销量最好;

3. 假设房子面积在90平方米以下的为小户型,90～140平方米的为中户型,140平方米以上的为大户型。分析计算大、中、小户型房屋销售情况。

第 5 章

电商大数据分析与探索

任务 5.1 "女装电子商务评论"数据分析

情境导入

电子商务的飞速发展离不开计算机与大数据技术的不断革新与发展。如今每年都有新的电商平台涌现,为了在激烈的市场竞争中占领一席之地,电商企业需要拥有更多的用户数据、商品数据、订单数据等海量信息,通过分析这些数据得出准确的用户需求、销售趋势等信息,并将这些信息应用在企业管理、数据应用和服务创新中。大数据分析能够将原本庞杂的数据进行挖掘和精准分析,帮助电商企业优化资源配置、降低运营成本、制定更合理的销售策略,进而提升市场份额和竞争力。

对于线上商店,客户对商品的评论情况在商品的销量、商店的经营和发展中起到了重要的作用。

现有一份某女装线上商店围绕客户评论制作的"女装电子商务评论"数据集文档"Clothing-Reviews.csv"。该数据集字段及其含义如表 5-1 所示。因为这是真实的商业数据,所以它被匿名化了,数据集中对公司的引用被替换为"零售商"。现在需要帮助商家对此份数据进行分析,提出一些建设性意见,帮助商店更好地发展。

表 5-1 "女装电子商务评论"数据集字段及其含义

字 段	含 义
order_id	订单编号
clothing_id	服装编号
age	年龄
review_text	评论
rating	评级
recommended_IND	是否推荐

续表

字　段	含　义
positive_feedback_count	积极反馈计数
class_name	服装分类

学习目标和要求

知识与技能目标

1. 掌握从不同数据源中创建 DataFrame 的方法。
2. 掌握操作 DataFrame 进行数据清洗的基本方法。
3. 掌握操作 DataFrame 进行数据转换的基本方法。
4. 掌握使用 Spark SQL 语句进行数据分析的方法。
5. 掌握持久存储数据的方法。

素质目标

1. 具有技术创新、与时俱进的开拓精神。
2. 具有脚踏实地的工匠精神和职业责任感。

5.1.1 数据准备

在实际的业务中，数据的来源多种多样。Spark SQL 在数据兼容方面，不仅可以直接处理 RDD，而且可以处理 Parquet 文件和 JSON 文件，以及外部数据库中的数据。Spark SQL 的一个重要特点是它能够统一处理关系表和 RDD，使数据分析人员可以轻松使用 SQL 命令进行外部查询，同时进行更复杂的数据分析。

在 Spark 中，DataFrame 是一种以 RDD 为基础的分布式数据集，类似传统数据库中的二维表格，是 Spark SQL 最核心的编程抽象。DataFrame 与 RDD 的主要区别在于，前者带有 schema 元信息，即 DataFrame 所表示的二维表数据集的每一列都带有名称和类型。这使 Spark SQL 得以洞察更多的结构信息，从而对 DataFrame 背后的数据源及作用于 DataFrame 上的变换进行针对性的优化，最终达到大幅提升运行效率的目标。

使用 Spark 库可以将不同数据来源的数据转换为 DataFrame，并对数据结果进行展示。创建 DataFrame 有很多种方法，比如从本地 List 中创建、从 RDD 中创建和从源数据中创建。下面简要介绍 3 种创建 DataFrame 的方法。

1. 使用 toDF() 函数创建 DataFrame

可以通过 toDF() 将本地序列（Seq）、数组或者 RDD 转换为 DataFrame，只要能对这些数据的内容指定数据类型即可。

1）使用本地序列和 toDF() 创建 DataFrame

首先，创建一个序列，将其定义为 df，内含服装序号、服装名称和服装价格 3 列；然后，使用 toDF() 指定这 3 列的名称，分别为 ID、

DataFrame
的创建方法

classification 和 Price。注意，如果直接使用 toDF() 函数而不指定列的名称，那么默认列名为"_1""_2""_3"。在序列创建完成后，使用 printSchema() 方法输出 DataFrame 的 schema，使用 show() 方法显示 DataFrame 中的内容。在 spark-shell 中输入以下命令，结果如图 5-1 所示。

```
// 创建序列并转换为 DataFrame
val df = Seq( (1, "Dresses", 100), (2, "Blouses", 50) ).toDF("ID", "classification", "Price")

// 输出 DataFrame 的 schema
df.printSchema()

// 显示数据内容
df.show()
```

```
scala> val df = Seq( (1, "Dresses", 100), (2, "Blouses", 50) )
.toDF("ID", "classification", "Price")
df: org.apache.spark.sql.DataFrame = [ID: int, classification:
 string ... 1 more field]

scala> df.printSchema()
root
 |-- ID: integer (nullable = false)
 |-- classification: string (nullable = true)
 |-- Price: integer (nullable = false)

scala> df.show()
+---+--------------+-----+
| ID|classification|Price|
+---+--------------+-----+
|  1|       Dresses|  100|
|  2|       Blouses|   50|
+---+--------------+-----+
```

图 5-1 使用本地序列和 toDF() 函数创建 DataFrame 示例

由图 5-1 可知，DataFrame 中 schema 的具体信息包含每一列的名称和数据类型，以及 DataFrame 中的详细数据。

2）使用 case 类和 toDF() 函数创建 DataFrame

利用反射机制推断 RDD 模式。使用这种方式创建 DataFrame，首先需要定义一个 case 类，因为只有 case 类才能被 Spark 隐式地转换为 DataFrame。

首先，定义一个名为 Goods 的 case 类，Goods 类包含 name 字符串和 price 整型数值两个字段；然后，利用 sc.textFile() 方法读取 HDFS 中的 "people.txt" 文件并创建 RDD，此文件需要提前上传到 HDFS 中；接着利用 map 转换操作，先引用 split("\t") 对数据进行分割，再引用 g => Goods(g(0), g(1).trim.toInt) 对 RDD 中的两列数据进行转换；最后利用 toDF() 函数，将数据转换为 DataFrame。

根据以上思路，在 spark-shell 中输入以下命令，创建 DataFrame，结果如图 5-2 所示。

```
// 使用 case 类定义模式
case class Goods(name: String, price: Int)

// 读取 HDFS 中的 "commodity.txt" 文件，得到一个 RDD 数据集，将该 RDD 数据集转换为 DataFrame
```

```
val good = sc.textFile("/Chapter5/commodity.txt").map(_.split("\t")).map(g => Goods(g(0), g(1).trim.toInt)).toDF()
```

// 输出 DataFrame 的 schema
```
good.printSchema()
```

// 显示数据内容
```
good.show()
```

```
scala> case class Goods(name: String, price: Int)
defined class Goods

scala> val good = sc.textFile("/Chapter5/commodity.txt").map(
_.split("\t")).map(g => Goods(g(0), g(1).trim.toInt)).toDF()
good: org.apache.spark.sql.DataFrame = [name: string, price: int]

scala> good.printSchema()
root
 |-- name: string (nullable = true)
 |-- price: integer (nullable = false)

scala> good.show()
+--------+-----+
|    name|price|
+--------+-----+
|    大衣|  100|
|    西装|  150|
|    围巾|  200|
|  运动衫|   50|
|  牛仔裤|  150|
|紧身女衫|  120|
|    短裤|   70|
|    夹克|  300|
+--------+-----+
```

图 5-2 使用 case 类和 toDF() 函数创建 DataFrame 示例

2. 使用 createDataFrame() 函数创建 DataFrame

在 SQLContext 中使用 createDataFrame() 函数也可以创建 DataFrame。和 toDF() 函数一样，这里创建的 DataFrame 的数据类型也可以是本地数组或者 RDD。首先创建一个 schema，然后用该 schema 指定一个 RDD，最后将此 RDD 和 schema 通过 createDataFrame() 函数转换为 DataFrame。

在通过 row 和 schema 创建 DataFrame 之前需要先导入相关的依赖包，再定义一个 schema，包含 IntegerType 类型的 id、StringType 类型的 name、IntegerType 类型的 price。通过 sc.parallelize() 方法对一个列表创建名称为 goodsRDD 的 RDD。最后将 schema 和 goodsRDD 通过 createDataFrame() 函数转换为 DataFrame。

在 spark-shell 中输入以下命令，结果如图 5-3 所示。

// 导入相关依赖包
```
import org.apache.spark.sql._
import org.apache.spark.sql.types._
```

// 指定一个 schema
```
val schema = StructType(
```

```
    List(
      StructField("id", IntegerType, true),
      StructField("name", StringType, true),
      StructField("price", IntegerType, true)
    )
  )
```

// 创建一个 RDD
var goodsRDD = sc.parallelize(List(Row(1," 裙子 ",300),Row(2, " 裤子 ", 225)))

// 将此 RDD 和 schema 通过 createDataFrame() 函数转换为 DataFrame
var goodsDF = spark.createDataFrame(goodsRDD, schema)

// 输出 DataFrame 的 schema
goodsDF.printSchema()

// 显示数据内容
goodsDF.show()

```
scala> import org.apache.spark.sql._
import org.apache.spark.sql._

scala> import org.apache.spark.sql.types._
import org.apache.spark.sql.types._

scala> val schema = StructType(
     |     List(
     |       StructField("id", IntegerType, true),
     |       StructField("name", StringType, true),
     |       StructField("price", IntegerType, true)
     |     )
     |   )
schema: org.apache.spark.sql.types.StructType = StructType(StructField(id,IntegerType,true), StructField(name,StringType,true), StructField(price,IntegerType,true))

scala> var goodsRDD = sc.parallelize(List(Row(1,"裙子",300),Row(2, "裤子", 225)))
goodsRDD: org.apache.spark.rdd.RDD[org.apache.spark.sql.Row] = ParallelCollectionRDD[52] at parallelize at <console>:30

scala> var goodsDF = spark.createDataFrame(goodsRDD, schema)
goodsDF: org.apache.spark.sql.DataFrame = [id: int, name: string ... 1 more field]

scala> goodsDF.printSchema()
root
 |-- id: integer (nullable = true)
 |-- name: string (nullable = true)
 |-- price: integer (nullable = true)

scala> goodsDF.show()
[Stage 4:>
+---+----+-----+
| id|name|price|
+---+----+-----+
|  1|裙子|  300|
|  2|裤子|  225|
+---+----+-----+
```

图 5-3　使用 createDataFrame() 函数创建 DataFrame 示例

3. 从外部数据中创建 DataFrame

1）使用 Parquet 文件创建 DataFrame

Parquet 是一种列式存储文件格式。作为 Apache 顶级项目被 Spark 吸收并作为默认数据源,在不指定读取和存储格式时,默认读写 Parquet 格式的文件。

假设读取 HDFS 中的"goods.parquet"文件,创建 DataFrame,命令如下:

```
val df = sqlContext.read.parquet("/Chapter5/goods.parquet")
```

2）使用 JSON 文件创建 DataFrame

JSON 是一种轻量级的数据交换格式,全称为 JavaScript Object Notation,是一种用规范的文本文件来存储和表示数据的格式。JSON 文件易于阅读和编写,同时易于机器的解析和生成,能有效提升网络传输效率。

假设读取 HDFS 中的"goods.json"文件,创建 DataFrame,命令如下:

```
val df = spark.read.json("/Chapter5/goods.json")
```

3）使用 CSV 文件创建 DataFrame

CSV（Comma-separated Values）,即逗号分隔值文件,以纯文本形式存储表格数据,有时也称为字符分隔值文件,因为分隔字符不只有逗号。CSV 文件由任意数目的记录组成,记录间以某种换行符分隔。每条记录由字段组成,字段间的分隔符是其他字符或者字符串,最常见的是逗号或者制表符。所有记录通常都有完全相同的字段序列。

使用 CSV 文件创建 DataFrame 有两种方式,一种是使用类型推断,另一种是自定义 schema。

（1）读取 CSV 文件,使用类型推断创建 DataFrame。

在 spark-shell 中输入以下命令,结果如图 5-4 所示。

```
// 定义 HDFS 中 "books.csv" 文件的路径
var file = "/Chapter5/books.csv"

// 加载路径,读取 CSV 文件,创建 DataFrame
var books = spark.read.option("header","true").csv(file)

// 打印 schema
books.printSchema()

// 打印 "books.csv" 文件中数据的条数
println(books.count())

// 显示数据内容
books.show()
```

```
scala> :paste
// Entering paste mode (ctrl-D to finish)

//定义HDFS中"books.csv"文件的路径
var file = "/Chapter5/books.csv"
//加载路径，读取CSV文件，创建DataFrame
var books = spark.read.option("header","true").csv(file)
//打印schema
books.printSchema()
//打印"books.csv"文件中数据的条数
println(books.count())
//显示数据内容
books.show()

// Exiting paste mode, now interpreting.

root
 |-- name: string (nullable = true)
 |-- author: string (nullable = true)
 |-- price: string (nullable = true)
 |-- publisher: string (nullable = true)

12
+----------+------+-----+------------------+
|      name|author|price|         publisher|
+----------+------+-----+------------------+
|    钱学森传|叶永烈|  null|              null|
|    三体全集|刘慈欣|   58|        重庆出版社|
|        活着|  余华|   31|  北京十月文艺出版社|
|        活着|  余华|   31|  北京十月文艺出版社|
|        null|  余华| null|              null|
|    千年一叹|余秋雨|   26|北京联合出版有限公司|
|        null|余秋雨|   26|              null|
|  万历十五年|黄仁宇| 40.5|          中华书局|
|  平凡的世界|  路遥| null|  北京十月文艺出版社|
|        null|  null| null|              null|
|  简读中国史|  null| null|              null|
|    千年一叹|余秋雨|   26|北京联合出版有限公司|
+----------+------+-----+------------------+

file: String = /Chapter5/books.csv
books: org.apache.spark.sql.DataFrame = [name: string, author: string ... 2 more fields]
```

图 5-4　使用类型推断创建 DataFrame

由图 5-4 中 schema 的内容可以看出，在将 schema 转换为 DataFrame 时，Spark SQL 会自动推断数据类型，所有字段均被设置成了 String 字符串类型。

（2）读取 CSV 文件，通过自定义 schema 创建 DataFrame。

通过自定义 schema 指定字段的数据类型，在 spark-shell 中输入以下命令，结果如图 5-5 所示。

```
import org.apache.spark.sql._
import org.apache.spark.sql.types._

// 定义 CSV 文件路径
var file = "/Chapter5/books.csv"

// 自定义一个 schema
var fields= List(
    StructField("book_name", StringType, true),
    StructField("book_author", StringType, true),
    StructField("price", DoubleType, true),
    )
val schema = StructType(fields)

// 读取 CSV 文件，并在创建 DataFrame 时指定 schema
```

```
var book = spark.read.option("header","true").schema(schema).csv(file)

// 打印 schema
book.printSchema()

// 显示数据内容
book.show()
```

```
scala> :paste
// Entering paste mode (ctrl-D to finish)

import org.apache.spark.sql._
import org.apache.spark.sql.types._
//定义CSV文件路径
var file = "/Chapter5/books.csv"
//自定义一个schema
var fields= List(
StructField("book_name", StringType, true),
StructField("book_author", StringType, true),
StructField("price", DoubleType, true),
     )
val schema = StructType(fields)
//读取CSV文件，并在创建DataFrame时指定schema
var book = spark.read.option("header","true").schema(schema).csv(file)
//打印schema
book.printSchema()
//显示数据内容
book.show()

// Exiting paste mode, now interpreting.

root
 |-- book_name: string (nullable = true)
 |-- book_author: string (nullable = true)
 |-- price: double (nullable = true)

2022-10-04 08:42:00,579 WARN csv.CSVHeaderChecker: Number of column in CSV header is not
equal to number of fields in the schema:
 Header length: 4, schema size: 3
CSV file: hdfs://master:8020/Chapter5/books.csv
+----------+-----------+-----+
| book_name|book_author|price|
+----------+-----------+-----+
|   钱学森传|     叶永烈| null|
|   三体全集|     刘慈欣| 58.0|
|       活着|       余华| 31.0|
|       活着|       余华| 31.0|
|       null|       余华| null|
|   千年一叹|     余秋雨| 26.0|
|       null|     余秋雨| 26.0|
| 万历十五年|     黄仁宇| 40.5|
| 平凡的世界|       路遥| null|
|       null|       null| null|
| 简读中国史|       null| null|
|   千年一叹|     余秋雨| 26.0|
+----------+-----------+-----+

import org.apache.spark.sql._
import org.apache.spark.sql.types._
file: String = /Chapter5/books.csv
fields: List[org.apache.spark.sql.types.StructField] = List(StructField(book_name,StringT
ype,true), StructField(book_author,StringType,true), StructField(price,DoubleType,true))
schema: org.apache.spark.sql.types.StructType = StructType(StructField(book_name,StringTy
pe,true), StructField(book_author,StringType,true), StructField(price,DoubleType,true))
book: org.apache.spark.sql.DataFrame = [book_name: string, book_author: string ... 1 more
 field]
```

图 5-5　通过自定义 schema 创建 DataFrame

由图 5-5 可知，通过自定义 schema 可以根据源数据情况将各字段数据定义成我们想要的数据类型，同时更改字段名称。在此示例中，book_name 和 book_author 字段的数据类型是 String，price 字段的数据类型是 DoubleType。通过自定义 schema 创建

DataFrame 的方式，可以让用户对处理的数据更熟悉。

4．读取"女装电子商务评论"数据集

本任务中的"女装电子商务评论"数据集是 CSV 格式的文件。首先将"Clothing-Reviews.csv"文件上传到 HDFS 中，然后加载数据集到 RDD 中，接着通过自定义 schema 的方式将 RDD 转换为 DataFrame，方便后续的数据探索与分析。

1）启动 Hadoop 集群和 Spark 集群

在 master 节点中输入以下命令，启动 Hadoop 集群，启动成功的结果如图 5-6 所示。

[root@master ~]# start-all.sh

```
[root@master ~]# start-all.sh
Starting namenodes on [master]
Last login: Tue Oct  4 02:07:39 EDT 2022 from 192.168.128.1 on pts/0
Starting datanodes
Last login: Tue Oct  4 02:07:55 EDT 2022 on pts/0
Starting secondary namenodes [master]
Last login: Tue Oct  4 02:07:58 EDT 2022 on pts/0
Starting resourcemanager
Last login: Tue Oct  4 02:08:10 EDT 2022 on pts/0
Starting nodemanagers
Last login: Tue Oct  4 02:08:22 EDT 2022 on pts/0
[root@master ~]#
```

图 5-6 启动 Hadoop 集群

在 master 节点中输入以下命令，启动 Spark 集群，启动成功的结果如图 5-7 所示。

[root@master ~]# start-spark-all.sh

```
[root@master ~]# start-spark-all.sh
starting org.apache.spark.deploy.master.Master, logging to /usr/local/src/spark/logs/spark-root-org.apache.spark.deploy.m
aster.Master-1-master.out
slave01: starting org.apache.spark.deploy.worker.Worker, logging to /usr/local/src/spark/logs/spark-root-org.apache.spark
.deploy.worker.Worker-1-slave01.out
slave02: starting org.apache.spark.deploy.worker.Worker, logging to /usr/local/src/spark/logs/spark-root-org.apache.spark
.deploy.worker.Worker-1-slave02.out
localhost: starting org.apache.spark.deploy.worker.Worker, logging to /usr/local/src/spark/logs/spark-root-org.apache.spa
rk.deploy.worker.Worker-1-master.out
[root@master ~]#
```

图 5-7 启动 Spark 集群

2）查看以上服务进程是否已经启动

在 master 节点中输入以下命令，查看服务进程。

[root@master ~]# jps

若正常显示图 5-8 所示的进程，则表示各项服务进程启动正常。

```
[root@master ~]# jps
1857 DataNode
2338 ResourceManager
2069 SecondaryNameNode
2902 Worker
1719 NameNode
2473 NodeManager
2842 Master
2973 Jps
```

图 5-8 查看服务进程

3）将"Clothing-Reviews.csv"文件上传到 HDFS 的"/Chapter5/"目录下

（1）将数据集上传到 Linux 本地"/root/data/Chapter5"目录下。通过 MobaXterm 工

具，直接将数据集拖动到"/root/data/Chapter5"目录下，如图 5-9 所示。

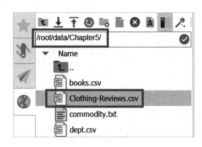

图 5-9　将数据集上传到 Linux 本地目录下

（2）在 master 节点中输入以下命令，将"/root/data/Chapter5"目录下的"Clothing-Reviews.csv"文件上传到 HDFS 中。

[root@master ~]# hdfs dfs -mkdir -p /Chapter5
[root@master ~]# hdfs dfs -put /root/data/Chapter5/Clothing-Reviews.csv /Chapter5

（3）在 master 节点中输入以下命令，查看该数据集是否已经上传到 HDFS 中，结果如图 5-10 所示，表示上传成功。

[root@master ~]# hdfs dfs -ls /Chapter5

```
[root@master ~]# hdfs dfs -ls /Chapter5
Found 1 items
-rw-r--r--   3 root supergroup    8272361 2022-10-04 02:52 /Chapter5/Clothing-Reviews.csv
[root@master ~]#
```

图 5-10　查看文件是否上传到 HDFS 中

4）读取 CSV 文件，通过自定义 schema 的方式创建 DataFrame

（1）在 master 节点中输入以下命令，启动 spark-shell，结果如图 5-11 所示。

[root@master ~]# spark-shell

```
[root@master ~]# spark-shell
2022-10-04 03:25:48,230 WARN util.NativeCodeLoader: Unable to load native-hadoop library for your platform... using built-in-java classes where applicable
Setting default log level to "WARN".
To adjust logging level use sc.setLogLevel(newLevel). For SparkR, use setLogLevel(newLevel).
Spark context Web UI available at http://master:4040
Spark context available as 'sc' (master = local[*], app id = local-1664868366749).
Spark session available as 'spark'.
Welcome to
      ____              __
     / __/__  ___ _____/ /__
    _\ \/ _ \/ _ `/ __/  '_/
   /___/ .__/\_,_/_/ /_/\_\   version 3.1.2
      /_/

Using Scala version 2.12.10 (Java HotSpot(TM) 64-Bit Server VM, Java 1.8.0_281)
Type in expressions to have them evaluated.
Type :help for more information.

scala>
```

图 5-11　启动 spark-shell

（2）读取 CSV 文件，通过自定义 schema 将其转换为 DataFrame。在 spark-shell 中输入以下命令，结果如图 5-12 所示。

import org.apache.spark.sql._

```
import org.apache.spark.sql.types._

// 读取 CSV 文件，自定义 schema
var file = "/Chapter5/Clothing-Reviews.csv"

// 指定一个 schema
var fields= List(
    StructField("order_id", IntegerType, true),
    StructField("clothing_id", StringType, true),
    StructField("age", IntegerType, true),
    StructField("review_text",StringType, true),
    StructField("rating",IntegerType, true),
    StructField("recommended_IND",StringType, true),
    StructField("positive_feedback_count",IntegerType, true),
    StructField("class_name",StringType, true),
    )
val schema = StructType(fields)

// 读取 CSV 文件并在创建 DataFrame 时指定 schema
var reviews = spark.read.option("header","true").schema(schema).csv(file)

// 打印 schema
reviews.printSchema()

// 显示数据内容
reviews.show()
```

```
scala> :paste
// Entering paste mode (ctrl-D to finish)

import org.apache.spark.sql._
import org.apache. park.sql.types._

//读取CSV文件，自定义schema
var file = "/Chapter5/Clothing-Reviews.csv"

//指定一个schema
var fields= List(
StructField("order_id", IntegerType, true),
StructField("clothing_id", StringType, true),
StructField("age", IntegerType, true),
StructField("review_text",StringType, true),
StructField("rating",IntegerType, true),
StructField("recommended_IND",StringType, true),
StructField("positive_feedback_count",IntegerType, true),
StructField("class_name",StringType, true),
    )
val schema = StructType(fields)

//读取CSV文件并在创建DataFrame时指定schema
var reviews = spark.read.option("header","true").schema(schema).csv(file)

//打印schema
reviews.printSchema()

//显示数据内容
reviews.show()

// Exiting paste mode, now interpreting.
```

图 5-12　读取 CSV 文件并将其转换为 DataFrame

```
root
 |-- order_id: integer (nullable = true)
 |-- clothing_id: string (nullable = true)
 |-- age: integer (nullable = true)
 |-- review_text: string (nullable = true)
 |-- rating: integer (nullable = true)
 |-- recommended_IND: string (nullable = true)
 |-- positive_feedback_count: integer (nullable = true)
 |-- class_name: string (nullable = true)

+--------+-----------+---+--------------------+------+---------------+-----------------------+----------+
|order_id|clothing_id|age|         review_text|rating|recommended_IND|positive_feedback_count|class_name|
+--------+-----------+---+--------------------+------+---------------+-----------------------+----------+
|       0|        767| 33|                null|     4|              1|                      0|  Intimates|
|       1|       1080| 34|                null|     5|              1|                      4|   Dresses|
|       2|       1077| 60| Some major design...|    3|              0|                      0|   Dresses|
|       2|       1077| 60| Some major design...|    3|              0|                      0|   Dresses|
|       3|       1049| 50|    My favorite buy!|     5|              1|                      0|     Pants|
|       4|        847| 47|    Flattering shirt|     5|              1|                      6|   Blouses|
|       5|       1080| 49|Not for the very ...|     2|              0|                      4|   Dresses|
|       6|        858| 39|Cagrcoal shimmer fun|     5|              1|                      1|     Knits|
|       6|        858| 39|Cagrcoal shimmer fun|     5|              1|                      1|     Knits|
|       7|        858| 39|   Shimmer, surprisi.|    4|              1|                      4|     Knits|
|       8|       1077| 24|          Flattering|     5|              1|                      0|   Dresses|
|       9|       1077| 34|    Such a fun dress!|    5|              1|                      0|   Dresses|
|      10|       1077| 53|   Dress looks like .|    3|              0|                     14|   Dresses|
|      11|       1095| 39|                null|     5|              1|                      2|   Dresses|
|      12|       1095| 53|            Perfect!!!|    5|              1|                      2|   Dresses|
|      13|        767| 44|            Runs big|     5|              1|                      0|  Intimates|
|      14|       1077| 50| Pretty party dres...|    3|              1|                      1|   Dresses|
|      15|       1065| 47| Nice, but not for...|    4|              1|                      3|     Pants|
|      16|       1065| 34| You need to be at...|    3|              1|                      2|     Pants|
|      17|        853| 41|   Looks great with .|    5|              1|                      0|   Blouses|
+--------+-----------+---+--------------------+------+---------------+-----------------------+----------+
only showing top 20 rows

import org.apache.spark.sql._
import org.apache.spark.sql.types._
file: String = /Chapter5/Clothing-Reviews.csv
fields: List[org.apache.spark.sql.types.StructField] = List(StructField(order_id,IntegerType,true), StructField(cl
othing_id,StringType,true), StructField(age,IntegerType,true), StructField(review_text,StringType,true), StructFie
ld(rating,IntegerType,true), StructField(recommended_IND,StringType,true), StructField(positive_feedback_count,Int
egerType,true), StructField(class_name,StringType,true))
schema: org.apache.spark.sql.types.StructType = StructType(StructField(order_id,IntegerType,true), StructField(clo
thing_id,StringType,true), StructField(age,IntegerType,true), StructField(review_text,StringTy...
```

图 5-12　读取 CSV 文件并将其转换为 DataFrame（续）

5.1.2　数据清洗

DataFrame 的
数据清洗方法

数据清洗是对数据进行重新审查和校验的过程，目的在于删除重复信息、纠正存在的错误，并保证数据一致性。在"Clothing-Reviews.csv"文件中，既存在一些重复数据，也有部分缺失，因此需要我们对其进行数据清洗。

查看"Clothing-Reviews.csv"文件的前 10 行。在 spark-shell 中输入"reviews.show(10)"，前 10 行数据如图 5-13 所示。从图中可以看出，在"order_id"是 2 和 6 的数据行中有重复；在"order_id"是 0 和 1 的数据行中，"review_text"列有空数据 null。

这只是此文件的前 10 行，而它拥有两万多行数据，因此我们需要对数据进行清洗，方便后续分析。下面以图 5-14 所示的 DataFrame 对象 books 的数据表为例，对数据清洗相关的方法进行讲解。数据表中存在完全重复的数据行，同时存在很多空数据。

```
scala> reviews.show(10)
+--------+-----------+---+--------------------+------+---------------+---------------------+----------+
|order_id|clothing_id|age|         review_text|rating|recommended_IND|positive_feedback_count|class_name|
+--------+-----------+---+--------------------+------+---------------+---------------------+----------+
|       0|        767| 33|                null|     4|              1|                    0|  Intimates|
|       1|       1080| 34|                null|     5|              1|                    4|   Dresses|
|       2|       1077| 60|Some major design...|     3|              0|                    0|   Dresses|
|       2|       1077| 60|Some major design...|     3|              0|                    0|   Dresses|
|       3|       1049| 50|    My favorite buy!|     5|              1|                    0|     Pants|
|       4|        847| 47|    Flattering shirt|     5|              1|                    6|   Blouses|
|       5|       1080| 49|Not for the very ...|     2|              0|                    4|   Dresses|
|       6|        858| 39|Cagrcoal shimmer fun|     5|              1|                    1|     Knits|
|       6|        858| 39|Cagrcoal shimmer fun|     5|              1|                    1|     Knits|
|       7|        858| 39|Shimmer, surprisi...|     4|              1|                    4|     Knits|
+--------+-----------+---+--------------------+------+---------------+---------------------+----------+
only showing top 10 rows
```

图 5-13　查看 Clothing-Reviews.csv"文件的前 10 行数据

```
+----------+--------+-----+--------------------+
|      name|  author|price|           publisher|
+----------+--------+-----+--------------------+
|  钱学森传|  叶永烈| null|                null|
|  三体全集|  刘慈欣|   58|          重庆出版社|
|      活着|    余华|   31|    北京十月文艺出版社|
|      活着|    余华|   31|    北京十月文艺出版社|
|      null|    余华| null|                null|
|  千年一叹|  余秋雨|   26|  北京联合出版有限公司|
|      null|  余秋雨|   26|                null|
|万历十五年|  黄仁宇| 40.5|            中华书局|
|平凡的世界|    路遥| null|    北京十月文艺出版社|
|      null|    null| null|                null|
|简读中国史|    null| null|                null|
|  千年一叹|  余秋雨|   26|  北京联合出版有限公司|
+----------+--------+-----+--------------------+
```

图 5-14　DataFrame 对象 books 的数据表

下面先介绍在 Spark SQL 中进行数据清洗的基本方法，再将这些方法应用到"Clothing-Reviews.csv"文件中。

1. drop(cols)

drop(cols) 方法表示按照列名 cols 删除 DataFrame 中的列，并返回新的 DataFrame。此方法可以删除数据表中无用的或者不想要的数据列。

在 spark-shell 中输入以下命令，删除"publisher"列，结果如图 5-15 所示。

var books1 = books.drop("publisher").show()

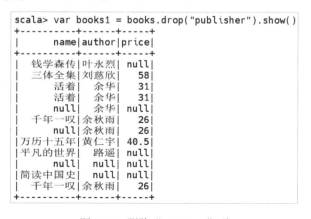

图 5-15　删除"publisher"列

2. dropDuplicates(subset=None)

dropDuplicates(subset=None) 方法用于删除 DataFrame 中的重复行，subset 用于指定在删除重复行时考虑的列。

（1）在 spark-shell 中输入以下命令，删除重复的行，结果如图 5-16 所示。

books.dropDuplicates().show()

由图 5-16 可知，在使用 dropDuplicates() 方法之后，数据表中完全重复的行已经不存在了。

（2）在 spark-shell 中输入以下命令，删除 "author" 列中重复的行，结果如图 5-17 所示。

books.dropDuplicates(List("author")).show()

图 5-16　删除 books 中重复的行

图 5-17　删除 "author" 列中重复的行

从图 5-14 中可以看出，在 "author" 列中，"余华""余秋雨""null" 这 3 个值存在重复。在应用 dropDuplicates(List("author")) 方法之后，从图 5-17 中可以看出，"author" 列的重复值没有了。

3. na.drop()

na.drop() 方法用于删除 DataFrame 中的空数据 null，可以通过加入 any 和 all 参数指定删除条件，加入数字参数指定有多少个空数据需要删除，加入字段名删除指定字段中的空数据。

（1）在 spark-shell 中输入以下命令，删除有空数据的行，结果如图 5-18 所示。

books.na.drop("any").show
books.na.drop().show　　// 此命令同上，在不指定参数时，与 any 的结果相同

由图 5-18 可以明显看出，在使用 na.drop() 方法处理之后，所有包含 null 的行全都被删除了。

（2）在 spark-shell 中输入以下命令，删除全部为空数据的行，结果如图 5-19 所示。

books.na.drop("all").show

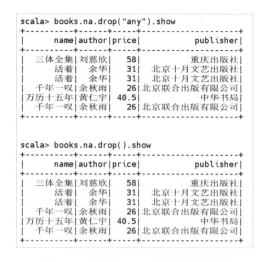

图 5-18　删除 books 中有空数据的行　　图 5-19　删除 books 中全部为空数据的行

由图 5-19 可知，在使用 na.drop("all") 方法处理之后，表格中 4 个字段均为 null 的那一行被删除了。

（3）在 spark-shell 中输入以下命令，删除有 3 个空数据的行，结果如图 5-20 所示。

books.na.drop(3).show()

由图 5-20 可知，表格的 4 个字段中有 3 个及以上为空数据的行被删除了。

（4）在 spark-shell 中输入以下命令，删除有两个及两个以上空数据的行，结果如图 5-21 所示。

books.na.drop(minNonNulls = books.columns.length - 2).show()

图 5-20　删除 books 中有 3 个空数据的行　　图 5-21　删除 books 中有两个及两个以上空数据的行

由图 5-21 可知，表格的 4 个字段中有两个及两个以上为空数据的行被删除了。

4．na.fill()

na.fill() 方法用于将 DataFrame 中所有的空数据填充为一个给定的字符串或数值，也可以为指定列中的空数据指定不同的字符串或数值。

（1）在 spark-shell 中输入以下命令，将表中所有的空数据都指定为"不明"，结果如图 5-22 所示。

books.na.fill(" 不明 ").show()

（2）在 spark-shell 中输入以下命令，对不同列中的空数据填充不同的字符串，结果如图 5-23 所示。

books.na.fill(Map("price"->"**", "publisher"->" 未知 ")).show()

图 5-22　将表中的空数据指定为 " 不明 "　　图 5-23　对不同列中的空数据填充不同的字符串

na.fill() 方法可以根据字段类型填充列值，这往往需要根据具体需求进行填补。

5. "女装电子商务评论" 数据集的数据清洗

对 "Clothing-Reviews.csv" 文件中的 "女装电子商务评论" 数据集进行数据清洗，删除重复的数据和有空数据的行。此过程基于 "5.1.1 数据准备" 中读取并转换为 DataFrame 的 reviews 进行操作。

（1）在 spark-shell 中输入以下命令，删除所有重复的行并统计数据量，结果如图 5-24 所示。

val reviews1=reviews.dropDuplicates()
// 查看处理 reviews 后的数据量
reviews1.count()

```
scala> val reviews1=reviews.dropDuplicates()
reviews1: org.apache.spark.sql.Dataset[org.apache.spark.sql.Row] = [orderid: string, Clothing ID: string ... 6 more fields]

scala> reviews1.count()
res62: Long = 23486
```

图 5-24　删除 "女装电子商务评论" 数据集中重复的行

（2）在 spark-shell 中输入以下命令，删除有空数据的行并统计数据量，结果如图 5-25 所示。

val reviews2=reviews1.na.drop()
// 统计处理 reviews1 后的数据量
reviews2.count()

```
scala> val reviews2=reviews1.na.drop()
reviews2: org.apache.spark.sql.DataFrame = [orderid: int, Clothing ID: string ... 6 more fields]

scala> reviews2.count()
res81: Long = 19660
```

图 5-25　删除 "女装电子商务评论" 数据集中有空数据的行

继续输入以下命令，查看删除重复行和所有有空数据的行之后的结果，如图 5-26 所示。

reviews2.show()

```
scala> reviews2.show()
+--------+-----------+---+--------------------+------+---------------+-----------------------+----------+
|order_id|clothing_id|age|         review_text|rating|recommended_IND|positive_feedback_count|class_name|
+--------+-----------+---+--------------------+------+---------------+-----------------------+----------+
|     260|       1115| 31|     Love this coat!|     5|              1|                      2| Outerwear|
|    1018|        936| 47|              Lovely|     5|              1|                      0|  Sweaters|
|    1927|       1059| 38|             So cute|     5|              1|                      3|     Pants|
|    2322|        992| 43|Too small, too co...|     1|              0|                      0|    Skirts|
|    2567|       1095| 36|              Beauty|     5|              1|                      3|   Dresses|
|    3081|        505| 26| Light and airy pjs|     5|              1|                      0|     Sleep|
|    3287|       1084| 42|Beautiful dress, ...|     3|              1|                      0|   Dresses|
|    3444|        939| 35|       Great jacket|     5|              1|                      2|  Sweaters|
|    3520|        872| 62|Loved so much i b...|     5|              1|                     11|     Knits|
|    4000|        860| 39|           Cute top!|     5|              1|                      3|     Knits|
|    4103|       1098| 35|    Didn't fit right|     2|              0|                      1|   Dresses|
|    4187|        860| 53|     Sweet and comfy|     4|              1|                      0|     Knits|
|    4275|       1086| 33|Perfect for year-...|     5|              1|                      0|   Dresses|
|    4579|        829| 58|             Love it|     5|              1|                      3|   Blouses|
|    5059|       1095| 65|    Summery-est ever|     5|              1|                      2|   Dresses|
|    5559|        877| 44|          Easy wear!|     5|              1|                      8|     Knits|
|    5592|        877| 51|Adds a little som...|     5|              1|                      0|     Knits|
|    5796|        921| 33|Beautiful piece; ...|     5|              1|                      0|  Sweaters|
|    6044|        895| 41|Why don't you loo...|     4|              1|                      0|Fine gauge|
|    6618|        863| 41|Cute - i ordered ...|     3|              1|                      2|     Knits|
+--------+-----------+---+--------------------+------+---------------+-----------------------+----------+
only showing top 20 rows
```

图 5-26 "女装电子商务评论"数据集清洗结果

由图 5-26 可知,通过删除重复的数据和有空数据的行,数据集中的数据量由原先的 23490 条,减少为 19660 条。

5.1.3 数据转换

DataFrame 的数据转换方法

数据转换是将数据从一种格式或结构转换为另一种格式或结构的过程,对于数据集成和数据管理等活动至关重要。它包括一系列活动:转换数据类型、丰富数据或执行聚合,具体取决于项目需求。

当数据分散存储在 RDD、JSON 文件和 SQL 数据库中时,我们要想将这些数据加载到 DataFrame 中,首先需要将它们全部合并,以便将数据作为一个整体进行清理、格式化,再转换为分析所需的格式。继续以图 5-4 中的数据为例,对数据转换的方法进行讲解。

1. withColumn(colName,col)

withColumn(colName,col) 方法用于为数据表添加新列,并返回一个新的 DataFrame。如果原来本身就有"colName"列,则进行替换。

在 spark-shell 中输入以下命令,为数据表添加或者替换新列,结果如图 5-27 所示。

```
// 添加"price"列,数值为原先数值的 2 倍,price 本身只要存在就进行替换
books.withColumn("price",$"price"*2).show()
// 添加"price2"新列
books.withColumn("price2",$"price"*2).show()
```

```
scala> books.withColumn("price",$"price"*2).show()
+----------+------+-----+------------------+
|      name|author|price|         publisher|
+----------+------+-----+------------------+
|   钱学森传|叶永烈| null|              null|
|   三体全集|刘慈欣|116.0|        重庆出版社|
|      活着|  余华| 62.0|  北京十月文艺出版社|
|      活着|  余华| 62.0|  北京十月文艺出版社|
|      null|  余华| null|              null|
|   千年一叹|余秋雨| 52.0|北京联合出版有限公司|
|      null|余秋雨| 52.0|              null|
|万历十五年|黄仁宇| 81.0|          中华书局|
|平凡的世界|  路遥| null|  北京十月文艺出版社|
|      null|  null| null|              null|
| 简读中国史|  null| null|              null|
|   千年一叹|余秋雨| 52.0|北京联合出版有限公司|
+----------+------+-----+------------------+

scala> books.withColumn("price2",$"price"*2).show()
+----------+------+-----+------------------+------+
|      name|author|price|         publisher|price2|
+----------+------+-----+------------------+------+
|   钱学森传|叶永烈| null|              null|  null|
|   三体全集|刘慈欣|   58|        重庆出版社| 116.0|
|      活着|  余华|   31|  北京十月文艺出版社|  62.0|
|      活着|  余华|   31|  北京十月文艺出版社|  62.0|
|      null|  余华| null|              null|  null|
|   千年一叹|余秋雨|   26|北京联合出版有限公司|  52.0|
|      null|余秋雨|   26|              null|  52.0|
|万历十五年|黄仁宇| 40.5|          中华书局|  81.0|
|平凡的世界|  路遥| null|  北京十月文艺出版社|  null|
|      null|  null| null|              null|  null|
| 简读中国史|  null| null|              null|  null|
|   千年一叹|余秋雨|   26|北京联合出版有限公司|  52.0|
+----------+------+-----+------------------+------+
```

图 5-27　为 books 添加 "price" 列

由图 5-27 可知，当表中已经存在 "price" 列时，如果替换的新列名也是 price，原来的 "price" 列就会被替换。如果替换的新列名是 price2，则原来的列会被保留，新列会被添加补充在表后。

2. withColumnRename(existing,new)

withColumnRename(existing,new) 方法用于对已经存在的列进行重命名，existing 为原来的列名，new 为新的列名，若名称不存在则这个操作不做任何事情。

在 spark-shell 中输入以下命令，修改 "publisher" 列的名称，结果如图 5-28 所示。

books.withColumnRenamed("publisher"," 出版社 ").show()

```
scala> books.withColumnRenamed("publisher","出版社").show()
+----------+------+-----+------------------+
|      name|author|price|            出版社|
+----------+------+-----+------------------+
|   钱学森传|叶永烈| null|              null|
|   三体全集|刘慈欣|   58|        重庆出版社|
|      活着|  余华|   31|  北京十月文艺出版社|
|      活着|  余华|   31|  北京十月文艺出版社|
|      null|  余华| null|              null|
|   千年一叹|余秋雨|   26|北京联合出版有限公司|
|      null|余秋雨|   26|              null|
|万历十五年|黄仁宇| 40.5|          中华书局|
|平凡的世界|  路遥| null|  北京十月文艺出版社|
|      null|  null| null|              null|
| 简读中国史|  null| null|              null|
|   千年一叹|余秋雨|   26|北京联合出版有限公司|
+----------+------+-----+------------------+
```

图 5-28　将 books 中的 "publisher" 列重命名

3. 合并数据表中的两列

UDF 是指用户定义函数,用于扩展系统的内置功能。我们可以通过在 Spark SQL 中自定义实际需要的 UDF 来处理数据。

首先创建连接方式,然后对 DataFrame 应用该连接方式,将两列合并成一列,方便数据的观察和信息的提取。

在 spark-shell 中输入以下命令,将 "name" 列和 "author" 列合并为一列,命名为 bookname-author,操作过程与结果如图 5-29 所示。

```
// 创建一个 UDF 来合并两列
val connect = udf((name: String, author: String) => {name + "-" + author})

// 应用 UDF 创建合并的列,并将其添加为新的列
val connectdf = books.withColumn("bookname-author",connect($"name", $"author"))

// 显示数据内容
connectdf.show
```

图 5-29 将 "name" 列和 "author" 列合并为一列

4. 为表中的列添加字符或者数字

在 spark-shell 中输入以下命令,将 "price" 列的值统一加上 10,结果如图 5-30 所示。

```
// 定义一个 UDF,将 "price" 列的值增加 10,此处需要利用 toDouble 将 price 值从 string 类型转换为 Double 类型
val addconst = udf((price:String) => {if(price==null) 0 else price.toDouble + 10})
// 利用 addconst() 方法将 "price" 列的值增加 10,并将 "price" 列重命名为 "价格"
val newbooks = books.withColumn(" 价格 ",addconst(col("price")))
newbooks.show
```

```
scala> val addconst = udf((price:String) => {if(price==null) 0 else price.toDouble + 10})
addconst: org.apache.spark.sql.expressions.UserDefinedFunction = SparkUserDefinedFunction($L
ambda$3740/275346701@23fab048,DoubleType,List(Some(class[value[0]: string])),Some(class[valu
e[0]: double]),None,false,true)

scala> val newbooks = books.withColumn("价格",addconst(col("price")))
newbooks: org.apache.spark.sql.DataFrame = [name: string, author: string ... 3 more fields]

scala> newbooks.show
+----------+------+-----+------------------+----+
|      name|author|price|         publisher|价格|
+----------+------+-----+------------------+----+
|   钱学森传|叶永烈|  null|              null| 0.0|
|   三体全集|刘慈欣|    58|        重庆出版社|68.0|
|       活着|  余华|    31|  北京十月文艺出版社|41.0|
|       活着|  余华|    31|  北京十月文艺出版社|41.0|
|       null|  余华|  null|              null| 0.0|
|   千年一叹|余秋雨|    26|北京联合出版有限公司|36.0|
|       null|余秋雨|    26|              null| 0.0|
|万历十五年|黄仁宇|  40.5|        中华书局|50.5|
|平凡的世界|  路遥|  null|  北京十月文艺出版社| 0.0|
|       null|  null|  null|              null| 0.0|
|   简读中国史|  null|  null|              null| 0.0|
|   千年一叹|余秋雨|    26|北京联合出版有限公司|36.0|
+----------+------+-----+------------------+----+
```

图 5-30 将"price"列中的数值统一加上 10

当对表中已经存在的列追加新的列时，如果新列与旧列同名，则会覆盖旧列。接着输入以下命令，结果如图 5-31 所示。

books.withColumn("price",addconst(col("price"))).show

```
scala> books.withColumn("price",addconst(col("price"))).show
+----------+------+-----+------------------+
|      name|author|price|         publisher|
+----------+------+-----+------------------+
|   钱学森传|叶永烈|  0.0|              null|
|   三体全集|刘慈欣| 68.0|        重庆出版社|
|       活着|  余华| 41.0|  北京十月文艺出版社|
|       活着|  余华| 41.0|  北京十月文艺出版社|
|       null|  余华|  0.0|              null|
|   千年一叹|余秋雨| 36.0|北京联合出版有限公司|
|       null|余秋雨| 36.0|              null|
|万历十五年|黄仁宇| 50.5|        中华书局|
|平凡的世界|  路遥|  0.0|  北京十月文艺出版社|
|       null|  null|  0.0|              null|
|   简读中国史|  null|  0.0|              null|
|   千年一叹|余秋雨| 36.0|北京联合出版有限公司|
+----------+------+-----+------------------+
```

图 5-31 对 books 追加"price"新列

5. na.replace()

na.replace() 方法可用于删除或替换现有的字符和数值。在处理数据的时候，很多时候会遇到批量替换的情况，如果逐个修改则效率过低且容易出错。na.replace() 方法为此提供了极大的便利。

在 spark-shell 中输入以下命令，将"publisher"列中的"北京十月文艺出版社"替换为"十月文艺"，结果如图 5-32 所示。（注意：如果 replace() 中的列名参数是"*"，那么将会对所有的列进行相应的替换。）

books.na.replace("publisher",Map(" 北京十月文艺出版社 " -> " 十月文艺 ")).show()

```
scala> books.na.replace("publisher",Map("北京十月文艺出版社" -> "十月文艺")).show()
+----------+------+-----+----------------+
|      name|author|price|       publisher|
+----------+------+-----+----------------+
|   钱学森传|叶永烈| null|            null|
|  三体全集|刘慈欣|   58|      重庆出版社|
|      活着|  余华|   31|        十月文艺|
|      活着|  余华|   31|        十月文艺|
|      null|  余华| null|            null|
|  千年一叹|余秋雨|   26|北京联合出版有限公司|
|      null|余秋雨|   26|            null|
|万历十五年|黄仁宇| 40.5|      中华书局|
|平凡的世界|  路遥| null|        十月文艺|
|      null|  null| null|            null|
|简读中国史|  null| null|            null|
|  千年一叹|余秋雨|   26|北京联合出版有限公司|
+----------+------+-----+----------------+
```

图 5-32　替换列中的字符串或数值

6. 数据转换

对"女装电子商务评论"数据集进行数据转换，将"recommended_IND（是否推荐）"列中的 1 替换为推荐，0 替换为不推荐。此转换基于上述数据清理之后的 reviews2 数据进行操作。

在 spark-shell 中输入以下命令，结果如图 5-33 所示。

val reviews3=reviews2.na.replace("recommended_IND",Map("0" -> " 不推荐 ","1" -> " 推荐 "))
reviews3.show

由图 5-33 可知，"recommended_IND"列中的 0 和 1 都被批量替换了。此处替换仅为了方便读者理解，实际工作中的数据表应尽量使用英文或者英文加数字，以提升数据在不同语言操作系统中的兼容性。

```
scala> val reviews3=reviews2.na.replace("recommended_IND",Map("0" -> "不推荐","1" -> "推荐"))
reviews3: org.apache.spark.sql.DataFrame = [order_id: int, clothing_id: string ... 6 more fields]

scala> reviews3.show
+--------+-----------+---+--------------------+------+---------------+---------------------+----------+
|order_id|clothing_id|age|         review_text|rating|recommended_IND|positive_feedback_count|class_name|
+--------+-----------+---+--------------------+------+---------------+---------------------+----------+
|     260|       1115| 31|      Love this coat!|     5|           推荐|                    2| Outerwear|
|    1018|        936| 47|              Lovely|     5|           推荐|                    0|  Sweaters|
|    1927|       1059| 38|             So cute|     5|           推荐|                    3|     Pants|
|    2322|        992| 43| Too small, too co...|     1|         不推荐|                    0|    Skirts|
|    2567|       1095| 36|              Beauty|     5|           推荐|                    3|   Dresses|
|    3081|        505| 26|  Light and airy pjs|     5|           推荐|                    0|     Sleep|
|    3287|       1084| 42| Beautiful dress, ...|     3|           推荐|                    0|   Dresses|
|    3444|        939| 35|        Great jacket|     5|           推荐|                    2|  Sweaters|
|    3520|        872| 62| Loved so much i b...|     5|           推荐|                   11|     Knits|
|    4000|        860| 39|            Cute top!|     5|           推荐|                    3|     Knits|
|    4103|       1098| 35|    Didn't fit right|     2|         不推荐|                    1|   Dresses|
|    4187|        860| 53|     Sweet and comfy|     4|           推荐|                    0|     Knits|
|    4275|       1086| 33| Perfect for year-...|     5|           推荐|                    0|   Dresses|
|    4579|        829| 58|             Love it|     5|           推荐|                    3|   Blouses|
|    5059|       1095| 65|    Summery-est ever|     5|           推荐|                    2|   Dresses|
|    5559|        877| 44|          Easy wear!|     5|           推荐|                    8|     Knits|
|    5592|        877| 51| Adds a little som...|     5|           推荐|                    0|     Knits|
|    5796|        921| 33| Beautiful piece; ...|     5|           推荐|                    0|  Sweaters|
|    6044|        895| 41| Why don't you loo...|     4|           推荐|                    0|Fine gauge|
|    6618|        863| 41| Cute - i ordered ...|     3|           推荐|                    2|     Knits|
+--------+-----------+---+--------------------+------+---------------+---------------------+----------+
only showing top 20 rows
```

图 5-33　"女装电子商务评论"数据集的数据转换

5.1.4 数据分析

Spark SQL
数据分析

使用 SQL 语句对数据进行探索分析是大多数软件开发和维护人员比较习惯的方式。Spark SQL 支持直接应用标准 SQL 语句进行查询。在进行 SQL 操作之前，需要先新建注册临时表，再进行 Spark SQL 查询。

下面以"books.csv"文件为例进行讲解。

1. 数据清洗

自定义 schema，读取"books.csv"文件，删除有空数据的行及重复数据。在 spark-shell 的 paste 模式下输入以下命令，结果如图 5-34 所示。

```
// 导入相关依赖包
import org.apache.spark.sql._
import org.apache.spark.sql.types._

// 定义 CSV 文件路径
var file = "/Chapter5/books.csv"

// 自定义一个 schema
var fields= List(
StructField("book_name", StringType, true),
StructField("book_author", StringType, true),
StructField("book_price", DoubleType, true),
StructField("book_publisher", StringType, true),
    )
val schema = StructType(fields)

// 读取 CSV 文件，在创建 DataFrame 时指定 schema
var book = spark.read.option("header","true").schema(schema).csv(file)

// 删除重复行
var book1 = book.dropDuplicates()
// 删除有空数据的行
var book2 = book1.na.drop("any")

// 打印 schema
book2.printSchema()

// 显示数据内容
book2.show()
```

```
scala> :paste
// Entering paste mode (ctrl-D to finish)

import org.apache.spark.sql._
import org.apache.spark.sql.types._

//定义CSV文件路径
var file = "/Chapter5/books.csv"

//自定义一个schema
var fields= List(
StructField("book_name", StringType, true),
StructField("book_author", StringType, true),
StructField("book_price", DoubleType, true),
StructField("book_publisher", StringType, true),
)
val schema = StructType(fields)

//读取CSV文件,在创建DataFrame时指定schema
var book = spark.read.option("header","true").schema(schema).csv(file)

//删除重复行
var book1 = book.dropDuplicates()
//删除有空数据的行
var book2 = book1.na.drop("any")

//打印schema
book2.printSchema()

//显示数据内容
book2.show()

// Exiting paste mode, now interpreting.

root
 |-- book_name: string (nullable = true)
 |-- book_author: string (nullable = true)
 |-- book_price: double (nullable = true)
 |-- book_publisher: string (nullable = true)

2022-10-05 10:59:25,331 WARN csv.CSVHeaderChecker: CSV header does not conform to the schema.
 Header: name, author, price, publisher
 Schema: book_name, book_author, book_price, book_publisher
Expected: book_name but found: name
CSV file: hdfs://master:8020/Chapter5/books.csv
+---------+-----------+----------+--------------------+
|book_name|book_author|book_price|      book_publisher|
+---------+-----------+----------+--------------------+
| 三体全集|     刘慈欣|      58.0|          重庆出版社|
|     活着|       余华|      31.0|    北京十月文艺出版社|
|万历十五年|     黄仁宇|      40.5|            中华书局|
| 千年一叹|     余秋雨|      26.0|    北京联合出版有限公司|
+---------+-----------+----------+--------------------+

import org.apache.spark.sql._
import org.apache.spark.sql.types._
file: String = /Chapter5/books.csv
fields: List[org.apache.spark.sql.types.StructField] = List(StructField(book_name,StringType,true), StructField(book_author,StringType,true), StructField(book_price,DoubleType,true), StructField(book_publisher,StringType,true))
schema: org.apache.spark.sql.types.StructType = StructType(StructField(book_name,StringType,true), StructField(book_author,StringType,true), StructField(book_price,DoubleType,true), StructField(book_publisher,StringType,true))
book: org.apache.spark.sql.DataFrame = [book_name: string, book_author: string ... 2 more fields]
book1: org.apache...
```

图 5-34 对"books.csv"文件进行数据清洗

2. 注册临时表

1) createGlobalTempView(name)

createGlobalTempView(name) 方法用于为 DataFrame 创建一个全局临时表。该临时表的生命周期和启动应用的周期一致,即只要启动的 Spark 应用存在,这个临时表就能一直被访问,直到退出应用为止。name 为新视图的名称,不能与已存在的视图名称一样,否则会报错。

在 spark-shell 中输入以下命令,为 book 创建全局临时表,结果如图 5-35 所示。

```
book2.createGlobalTempView("bglobal")
// 查询 bglobal 临时表
spark.sql("select * from global_temp.bglobal").show()
```

```
scala> book2.createGlobalTempView("bglobal")

scala> spark.sql("select * from global_temp.bglobal").show()
2022-10-05 11:02:30,470 WARN csv.CSVHeaderChecker: CSV header does not conform to the schema.
 Header: name, author, price, publisher
 Schema: book_name, book_author, book_price, book_publisher
Expected: book_name but found: name
CSV file: hdfs://master:8020/Chapter5/books.csv
+---------+-----------+----------+--------------------+
|book_name|book_author|book_price|      book_publisher|
+---------+-----------+----------+--------------------+
|   三体全集|       刘慈欣|      58.0|            重庆出版社|
|      活着|         余华|      31.0|      北京十月文艺出版社|
| 万历十五年|       黄仁宇|      40.5|              中华书局|
|   千年一叹|       余秋雨|      26.0|      北京联合出版有限公司|
+---------+-----------+----------+--------------------+
```

图 5-35 使用 createGlobalTempView(name) 方法创建表

2）createOrReplaceGlobalTempView(name)

createOrReplaceGlobalTempView(name) 方法用于创建或替换视图。当视图名称已经存在时，则进行替换；若其不存在，则进行创建。在 spark-shell 中输入以下命令，结果如图 5-36 所示。

// 创建全局临时视图，若 name 已存在则覆盖，若不存在则进行创建
book2.createOrReplaceGlobalTempView("bglobal")
// 查询视图
spark.sql("select * from global_temp.bglobal").show()

```
scala> book2.createOrReplaceGlobalTempView("bglobal")

scala> spark.sql("select * from global_temp.bglobal").show()
2022-10-05 11:03:04,090 WARN csv.CSVHeaderChecker: CSV header does not conform to the schema.
 Header: name, author, price, publisher
 Schema: book_name, book_author, book_price, book_publisher
Expected: book_name but found: name
CSV file: hdfs://master:8020/Chapter5/books.csv
+---------+-----------+----------+--------------------+
|book_name|book_author|book_price|      book_publisher|
+---------+-----------+----------+--------------------+
|   三体全集|       刘慈欣|      58.0|            重庆出版社|
|      活着|         余华|      31.0|      北京十月文艺出版社|
| 万历十五年|       黄仁宇|      40.5|              中华书局|
|   千年一叹|       余秋雨|      26.0|      北京联合出版有限公司|
+---------+-----------+----------+--------------------+
```

图 5-36 使用 createOrReplaceGlobalTempView(name) 方法创建视图

3）createTempView(name)

createTempView(name) 方法用于创建临时视图。name 为新视图的名称，名称不能与已存在的视图名称一样，否则会报错。在 spark-shell 中输入以下命令，结果如图 5-37 所示。

// 创建名为 btemp 的视图
book2.createTempView("btemp")
// 查询视图
spark.sql("select * from btemp").show()

```
scala> book2.createTempView("btemp")

scala> spark.sql("select * from btemp").show()
2022-10-05 11:03:49,355 WARN csv.CSVHeaderChecker: CSV header does not conform to the schema.
 Header: name, author, price, publisher
 Schema: book_name, book_author, book_price, book_publisher
Expected: book_name but found: name
CSV file: hdfs://master:8020/Chapter5/books.csv
+---------+-----------+----------+--------------------+
|book_name|book_author|book_price|      book_publisher|
+---------+-----------+----------+--------------------+
|   三体全集|       刘慈欣|      58.0|            重庆出版社|
|      活着|         余华|      31.0|      北京十月文艺出版社|
| 万历十五年|       黄仁宇|      40.5|              中华书局|
|   千年一叹|       余秋雨|      26.0|      北京联合出版有限公司|
+---------+-----------+----------+--------------------+
```

图 5-37 使用 createTempView(name) 方法创建视图

4) createOrReplaceTempView(name)

createOrReplaceTempView(name) 方法用于为 DataFrame 创建本地的临时视图，其生命周期只限于当前的 SparkSession。当视图名称已经存在时，则进行替换；若其不存在，则进行创建。在 spark-shell 中输入以下命令，结果如图 5-38 所示。

```
// 创建本地的临时视图
book2.createOrReplaceTempView("btemp")
// 查询视图
spark.sql("select * from btemp").show()
```

图 5-38　使用 createOrReplaceTempView(name) 方法创建视图

3. Spark SQL 查询

在临时表创建完成之后，就可以通过执行 SQL 语句对表进行查询了。在 spark-shell 中输入以下命令，进行简单的 SQL 查询操作，结果如图 5-39 所示。

```
// 在临时表"btemp"中执行 SQL 语句，查询整个表中的数据
var bookresult = spark.sql("select * from btemp")

// 显示 DF
bookresult.show

// 查询 book_price 小于 30 的书
spark.sql("select * from btemp where book_price<30").show
```

图 5-39　对临时表"btemp"进行简单的 SQL 查询操作

4. SQL 探索分析

对数据转换之后的"女装电子商务评论"DataFrame 对象 reviews3 创建临时视图，并进行 SQL 探索分析。

（1）在 spark-shell 中输入以下命令，使用 createOrReplaceTempView() 方法创建本地的临时视图"clothing_reviews"，结果如图 5-40 所示。

```
reviews3.createOrReplaceTempView("clothing_reviews")
```

```
scala> reviews3.createOrReplaceTempView("clothing_reviews")
scala>
```

图 5-40　对 reviews 创建临时视图

（2）在 spark-shell 中输入以下命令，查看 40 岁以下不同年龄段的客户人数。

在以下代码中，将统计的年龄数量 count(age) 重新命名为 total_nu，并将统计的结果按照年龄进行升序排序。

```
spark.sql("""select age,count(age) as total_nu
    from clothing_reviews
    where age<40
    group by age
    order by age""").show
```

```
scala> reviews3.createOrReplaceTempView("clothing_reviews")
scala> spark.sql("""select age,count(age) as total_nu
    |         from clothing_reviews
    |         where age<40
    |         group by age
    |         order by age""").show
+---+--------+
|age|total_nu|
+---+--------+
| 18|       4|
| 19|      28|
| 20|      91|
| 21|      75|
| 22|     106|
| 23|     210|
| 24|     201|
| 25|     276|
| 26|     369|
| 27|     300|
| 28|     358|
| 29|     446|
| 30|     343|
| 31|     477|
| 32|     511|
| 33|     606|
| 34|     669|
| 35|     728|
| 36|     663|
| 37|     626|
+---+--------+
only showing top 20 rows
```

图 5-41　查看 40 岁以下不同年龄段的客户人数

由图 5-41 可知，在 40 岁以下不同年龄段的客户中，以 26 岁到 37 岁居多，20 岁左右的客户较少。

（3）在 spark-shell 中输入以下命令，统计 recommended_IND 为"不推荐"的订单的评分分布。对"rating"列进行 count(1) 计数，并限定条件为 recommended_IND = '不

推荐 '，最后根据 rating 的值进行 group 分类，结果如图 5-42 所示。

```
spark.sql("""select rating, count(1)
    from clothing_reviews
    where recommended_IND = '不推荐'
    group by rating""").show
```

图 5-42　recommended_IND 为"不推荐"的订单的评分分布

由图 5-42 可知，不推荐的订单评分集中在 2 分和 3 分，4 分和 5 分中也有少量不推荐的订单。

（4）为进一步了解客户不推荐的原因，查看客户的具体评论。

在 spark-shell 中输入以下命令，查看评分小于或等于 3 且为"不推荐"的订单的具体评论。选择"clothing_id""rating""review_text"这 3 列进行显示，并限定条件为 rating<=3 且 recommended_IND = '不推荐'，结果如图 5-43 所示。

```
spark.sql("""select clothing_id,rating,review_text
    from clothing_reviews
    where rating<=3 and recommended_IND = '不推荐'""").show
```

图 5-43　评分小于或等于 3 且为"不推荐"的订单的具体评论

由图 5-43 可知，用户不推荐的具体理由有"服装过小""裤子质量不高""非常失望"等，可供商家查看。

（5）此外，还可以查询不推荐订单分别属于哪些服装分类，以及各种服装分类的不推荐情况。

在 spark-shell 中输入以下命令，查看不同服装分类的不推荐数量。对"class_name"列进行 count() 统计并重命名为 classnum，限定条件为 recommended_IND = '不推荐'，最后根据 class_name 的值分类，结果如图 5-44 所示。

```
spark.sql("""select class_name, count(1) as classnum
    from clothing_reviews
    where recommended_IND = '不推荐'
    group by class_name""").show
```

```
scala> spark.sql("""select class_name, count(1) as classnum
    | from clothing_reviews
    | where recommended_IND = '不推荐'
    | group by class_name""").show
+----------+--------+
|class_name|classnum|
+----------+--------+
|   Dresses|    1057|
| Outerwear|      50|
|      Swim|      60|
|   Blouses|     493|
|     Knits|     771|
|   Jackets|      90|
|     Trend|      26|
|    Lounge|      82|
|     Pants|     184|
|Fine gauge|     153|
|  Layering|      13|
|    Skirts|     121|
|  Intimates|     16|
|   Legwear|      20|
|     Sleep|      26|
|    Shorts|      44|
|  Sweaters|     249|
|     Jeans|     120|
+----------+--------+
```

图 5-44　不同服装分类的不推荐情况

在图 5-44 中可以查看不同服装分类的不推荐订单数量，如第一行的裙装类，不推荐的订单量达到了 1057 件。

（6）除了查询"不推荐"、"评分"低的订单情况，还可以统计分析不同评分等级的服装编号"clothing_id"。此处，服装编号代表具体某件服装。

在 spark-shell 中输入以下命令，获得不同服装编号"clothing_id"的五星好评数量，并且降序排序。对"clothing_id"列进行 count() 统计，重命名为 five_stars_num 并输出。限定条件为 rating=5，根据"clothing_id"进行分类，最后根据五星好评数量"five_stars_num"进行降序排序，结果如图 5-45 所示。

```
spark.sql("""select clothing_id,count(1) as five_stars_num
    from clothing_reviews
    where rating=5
    group by clothing_id
    order by five_stars_num desc""").show
```

第 5 章 电商大数据分析与探索

```
scala> spark.sql("""select clothing_id,count(1) as five_stars_num
     | from clothing_reviews
     | where rating=5
     | group by clothing_id
     | order by five_stars_num desc""").show
+-----------+--------------+
|clothing_id|five_stars_num|
+-----------+--------------+
|       1078|           467|
|        862|           376|
|       1094|           347|
|       1081|           283|
|        872|           269|
|       1110|           240|
|        829|           239|
|        895|           198|
|        936|           175|
|        850|           171|
|        868|           166|
|        867|           157|
|       1059|           154|
|       1080|           141|
|        863|           140|
|       1095|           138|
|       1083|           132|
|       1086|           131|
|       1077|           122|
|       1022|           121|
+-----------+--------------+
only showing top 20 rows
```

图 5-45　不同服装的五星好评数量

由图 5-45 可知，"clothing_id" 为 1078 的服装的五星好评数量最多。

（7）使用 Spark SQL 的内置函数进行相关性等分析。

① 分析 "rating" 评分字段和 "positive_feedback_count" 积极反馈数量字段之间的相关性。

在数据分析中，相关性分析可以用于分析两个变量之间的相关方向和相关强度，从而在改变其中一个变量时影响另一个变量的变化。

可以利用协方差函数 cov() 判断两个变量之间的相关方向。协方差表示两个变量的总体误差。如果两个变量的变化趋势一致，即其中一个变量大于自身的期望值，另一个变量也大于自身的期望值，那么两个变量之间的协方差就是正值。如果两个变量的变化趋势相反，即其中一个变量大于自身的期望值，另一个变量却小于自身的期望值，那么两个变量之间的协方差就是负值。

在 spark-shell 中输入以下命令，分析 DataFrame 对象 reviews3 中 "rating" 评分字段和 "positive_feedback_count" 积极反馈数量字段之间的相关方向，结果如图 5-46 所示。

```
//cov() 给出了这两个字段之间的相关方向
val cov = reviews3.stat.cov("rating","positive_feedback_count")
println("rating to positive_feedback_count : Covariance = %.4f".format(cov))
```

```
scala> val cov = reviews3.stat.cov("rating","positive_feedback_count")
cov: Double = -0.3677929631033124

scala> println("rating to positive_feedback_count : Covariance = %.4f".format(cov))
rating to positive_feedback_count : Covariance = -0.3678
```

图 5-46　两个字段之间的相关方向

可以利用相关性函数 corr() 计算两个变量之间的相关性。在 spark-shell 中输入以下命令，分析这两个字段之间的相关强度，结果如图 5-47 所示。

```
//corr() 给出了这两个字段之间的相关强度
val corr = reviews3.stat.corr("rating","positive_feedback_count")
println("rating to positive_feedback_count : Correlation = %.4f".format(corr))
```

```
scala> val corr = reviews3.stat.corr("rating","positive_feedback_count")
corr: Double = -0.05667590673474581

scala> println("rating to positive_feedback_count : Correlation = %.4f".format(corr))
rating to positive_feedback_count : Correlation = -0.0567
```

图 5-47　两个变量之间的相关强度

由图 5-47 可知，在此数据集中，评分高低和积极反馈数量的相关性不大。

② 使用 freqItems() 函数提取数据列中最常出现的项。

在数据分析中，根据业务需求，可能需要知晓数据列中最常出现的元素，从而为决策提供依据。可以利用 freqItems() 函数提取数据列中最常出现的项。freqItems() 函数中第 1 个参数是指定的分析列，第 2 个参数是指定的支持级别。在这里，我们选择 "class_name" 作为分析列，在 DataFrame reviews3 中，找到服装分类出现概率大于 40% 的类别，即支持级别为 0.4。

在 spark-shell 中输入以下命令，结果如图 5-48 所示。

```
val freq = reviews3.stat.freqItems(Seq("class_name"), 0.4)
freq.show(false)
```

```
scala> val freq = reviews3.stat.freqItems(Seq("class_name"), 0.4)
freq: org.apache.spark.sql.DataFrame = [class_name_freqItems: array<string>]

scala> freq.show(false)
+--------------------+
|class_name_freqItems|
+--------------------+
|[Knits, Dresses]    |
+--------------------+
```

图 5-48　提取数据列中最常出现的项

由图 5-48 可知，最常出现的服装分类是 "Knits" 针织衫和 "Dresses" 裙子。

③ 使用 approxQuantile() 函数计算 DataFrame 中数字列的近似分位数。

分位数表示在这个样本集中，在对数字列从小到大排列之后，小于某值的样本子集占总样本集的比例。

对 "age" 列计算近似分位数，指定的分位数概率为 0.25、0.5 和 0.75（注意：0 是最小值，1 是最大值，0.5 是中值（median）。

在 spark-shell 中输入以下命令，结果如图 5-49 所示。

```
val quantiles = reviews3.stat.approxQuantile("age", Array(0.25,0.5,0.75),0.0)
```

```
scala> val quantiles = reviews3.stat.approxQuantile("age", Array(0.25,0.5,0.75),0.0)
quantiles: Array[Double] = Array(34.0, 41.0, 52.0)
```

图 5-49　计算数字列的近似分位数结果

由图 5-49 可知，第 1 个四分位数是 34.0，中分位数是 41.0，第 3 个四分位数是 52.0，表示年龄小于 34 岁的样本子集占总样本集的 25%，小于 41 岁的样本子集占 50%，小于 52 岁的样本子集占 75%。

④ 使用 crosstab() 函数探索年龄"age"和评分"rating"之间是否有相关性。

crosstab() 函数可以对两个变量中的每个数据都进行相关性计算，从而观察这两个变量之间相互关系的程度与趋势。

在 spark-shell 中输入以下命令，结果如图 5-50 所示。

reviews3.stat.crosstab("age","rating").orderBy("age_rating").show()

```
scala> reviews3.stat.crosstab("age","rating").orderBy("age_rating").show()
+----------+---+---+---+---+---+
|age_rating|  1|  2|  3|  4|  5|
+----------+---+---+---+---+---+
|        18|  0|  0|  0|  2|  2|
|        19|  1|  0|  4| 13| 10|
|        20|  3|  3| 11| 16| 58|
|        21|  3|  3|  6| 15| 48|
|        22|  1|  3| 14| 19| 69|
|        23|  3| 13| 20| 55|119|
|        24|  5| 11| 24| 45|116|
|        25| 13| 16| 32| 63|152|
|        26|  9| 21| 68| 72|199|
|        27|  8| 21| 31| 71|169|
|        28| 23| 33| 53| 69|180|
|        29| 16| 32| 75| 99|224|
|        30| 16| 29| 52| 61|185|
|        31| 18| 33| 80|103|243|
|        32| 20| 30| 65|103|293|
|        33| 24| 45| 86|143|308|
|        34| 20| 71| 92|146|340|
|        35| 28| 71| 85|149|395|
|        36| 19| 46| 86|127|385|
|        37| 24| 50| 92|125|335|
+----------+---+---+---+---+---+
only showing top 20 rows
```

图 5-50　年龄与评分之间的相关性

从图 5-50 所示的结果中可以观察到年龄与评分之间的数量关系。

通过以上对"女装电子商务评论"数据集的分析，可以获得以下信息：不推荐的订单的评分集中在 2 分和 3 分，可以重点查看此评分的具体评论；通过不同服装分类的不推荐数量，可以获知哪些分类的问题较多，哪些分类比较受欢迎；通过统计五星好评数量，可以获知哪些服装比较畅销；通过各字段之间的相关性及函数分析，可以获取更细致的信息。这些信息都可以为商家提供决策依据，为商店的发展提供建议。

5.1.5　数据输出

对数据进行分析后的结果最终需要保存并输出，以供决策者进行查看和引用。DataFrame 提供了很多存储方法，可以存储到数据库中，也可以保存到文件中。将分析结果保存到文件中有两种方法，下面以将数据写入 HDFS 中为例进行详细讲解。

方法一：利用 save() 函数保存数据。

在 spark-shell 中输入以下命令，选择满足条件的数据，使用 write.format() 函数将其保存为 CSV 格式的文件，并上传到 HDFS 的"/Chapter5/reviews"目录中，结果如图 5-51 所示。

// 选取 review3 中的相关数据列进行保存

```
var selectData = reviews3.select("order_id","clothing_id","age","review_text","rating","recommended_
IND","positive_feedback_count","class_name")
```

// 调用 save() 函数，将 DataFrame 数据保存到 "/Chapter5/reviews" 目录中
selectData.select("*").write.format("csv").save("/Chapter5/reviews")

```
scala> var selectData = reviews3.select("order_id","clothing_id","age","revi
ew_text","rating","recommended_IND","positive_feedback_count","class_name")
selectData: org.apache.spark.sql.DataFrame = [order_id: int, clothing_id: st
ring ... 6 more fields]

scala> selectData.select("*").write.format("csv").save("/Chapter5/reviews")
```

图 5-51　存储 DataFrame 的方法（1）

在保存数据之后，返回 master 节点，在 master 节点中输入以下命令，输出保存的数据，如图 5-52 所示。

[root@master ~]#hdfs dfs -ls /Chapter5/reviews

```
[root@master ~]# hdfs dfs -ls /Chapter5/reviews
Found 201 items
-rw-r--r--   3 root supergroup          0 2022-10-06 04:07 /Chapter5/reviews/_SUCCESS
-rw-r--r--   3 root supergroup       5093 2022-10-06 04:07 /Chapter5/reviews/part-00000-c73bbeb8-8a5f-4e4a-bea4-39e7c5
6bce2e-c000.csv
```

图 5-52　在 HDFS 中查看保存的数据（1）

write.format() 函数支持输出 JSON、Parquet、JDBC、orc、CSV、text 等文件格式，save() 函数用于定义保存的位置。在保存成功之后，我们可以在保存位置的目录下看到保存结果，但是它并不是一个文件，而是一个目录。在读取的时候，并不需要使用文件夹里面的"part-"文件，直接读取目录即可。

方法二：先通过 df.rdd.saveAsTextFile("file:///") 将数据转化成 RDD 再保存。此方法可以将数据存储为 txt 格式。

在 spark-shell 中输入以下命令，将数据保存到 HDFS 的 "/Chapter5" 目录中，并重命名为 reviews_two.txt，结果如图 5-53 所示。

reviews3.rdd.saveAsTextFile("file:///Chapter5/reviews_two.txt")

```
scala> reviews3.rdd.saveAsTextFile("/Chapter5/reviews_two.txt")
scala>
```

图 5-53　存储 DataFrame 的方法（2）

在保存数据之后，返回 master 节点，继续输入以下命令，输出保存的数据，如图 5-54 所示。

[root@master ~]#hdfs dfs -ls /Chapter5/

第 5 章 电商大数据分析与探索

```
[root@master ~]# hdfs dfs -ls /Chapter5/
Found 5 items
-rw-r--r--   3 root supergroup    1048422 2022-10-05 23:15 /Chapter5/Clothing-Reviews.csv
-rw-r--r--   3 root supergroup        456 2022-10-04 10:23 /Chapter5/books.csv
-rw-r--r--   3 root supergroup        104 2022-10-04 08:04 /Chapter5/commodity.txt
drwxr-xr-x   - root supergroup          0 2022-10-06 04:07 /Chapter5/reviews
drwxr-xr-x   - root supergroup          0 2022-10-06 07:44 /Chapter5/reviews_two.txt
[root@master ~]#
```

图 5-54　在 HDFS 中查看保存的数据（2）

任务 5.2　"在线销售订单"数据分析

情境导入

电商大数据改变了我们的生活。除了精准营销，电商还可以依据客户消费习惯来提前为客户备货，并利用便利店作为货物中转点，在客户下单后 15 分钟内将货物送上门，提升客户体验。菜鸟网络宣称目标是在 24 小时内完成在中国境内的送货，以及京东宣传在下单后 15 分钟内上门取货都是基于客户消费习惯的大数据分析和预测。

同时，随着电商直播与购物的盛行，大型零售商正在积极寻找增加利润的方法。销售分析是大型零售商通过了解客户的购买行为和模式来增加销售额的关键技术之一，包括查找客户购买行为数据集，探索数据集上下文中组合在一起的项目之间的关系。现有两份"在线销售订单"数据集文件"orders.csv"和"order-details.csv"，包含不同地区、多种商品的销售情况，数据字段及其含义如表 5-2、表 5-3 所示。

表 5-2　"订单信息"数据集字段及含义

字　　段	含　　义
order_id	订单编号
order_date	订单日期
cust_id	客户编号
full_name	客户名字
city	城市

表 5-3　"订单详细信息"数据集字段及含义

字　　段	含　　义
order_id	订单编号
sku	商品的编码
qty_ordered	已订购数量
price	商品价格
category	类别
discount_percent	折扣百分比

现在希望通过数据分析了解以下情况。

1．每个客户的下单数量是多少？
2．每个城市的订单是多少？
3．每个客户的消费总额是多少？
4．客户的平均订单消费额是多少？

学习目标和要求

知识与技能目标
1. 能灵活使用 DataFrame 数据操作的各种方法进行数据查询。
2. 能使用聚合、连接等操作对 DataFrame 数据进行转换。
3. 能使用聚合函数对 DataFrame 数据进行分析。

素质目标
1. 具有不断探究的钻研精神。
2. 提升大数据分析工作岗位的职业荣誉感。

5.2.1 数据查询操作

在 Spark SQL 中，DataFrame 提供了很多查看和获取数据的操作方法。在对"在线销售订单"数据集进行分析和探索之前，本节将以"企业部门信息"数据集文件"dept.csv"和"员工个人信息"数据集文件"emp.csv"为例，介绍直接在 DataFrame 对象上进行查询操作的方法。

DataFrame
数据查询操作

将"企业部门信息"数据集文件"dept.csv"和"员工个人信息"数据集文件"emp.csv"上传到 HDFS 中，分别创建 DataFrame 对象 dept 和 emp。

在 spark-shell 中输入以下命令，option("header","true") 表示在创建 DataFrame 时将第一行作为数据表的字段名，option("inferSchema","true") 表示自动推断数据类型，结果如图 5-55 所示。

```
// 创建 DataFrame 对象 dept
var file1 = "/Chapter5/dept.csv"
val dept = spark.read.option("header","true").
    option("inferSchema","true").
    csv("/Chapter5/dept.csv")

// 创建 DataFrame 对象 emp
var file2 = "/Chapter5/emp.csv"
val emp = spark.read.option("header","true").
    option("inferSchema","true").
    csv("/Chapter5/emp.csv")
```

```
scala> var file1 = "/Chapter5/dept.csv"
file1: String = /Chapter5/dept.csv

scala> val dept = spark.read.option("header","true").
     |              option("inferSchema","true").
     |              csv("/Chapter5/dept.csv")
dept: org.apache.spark.sql.DataFrame = [dept_id: int, dept_name: string ... 1 more field]

scala> var file2 = "/Chapter5/emp.csv"
file2: String = /Chapter5/emp.csv

scala> val emp = spark.read.option("header","true").
     |              option("inferSchema","true").
     |              csv("/Chapter5/emp.csv")
emp: org.apache.spark.sql.DataFrame = [staff_id: int, name: string ... 4 more fields]
```

图 5-55 创建 DataFrame 对象 dept 和 emp

1. 查看数据

1）show()

在之前的示例中，已经多次用过 show() 方法，它主要用于测试，查看输入的数据、获得的结果是否正确，能够以表格的形式展示 DataFrame 中的数据。

（1）没有参数的 show()。

没有参数的 show() 方法默认最多能够显示 20 条数据。

在 spark-shell 中输入以下命令，显示 emp 和 dept 中的数据，结果如图 5-56、图 5-57 所示。

```
emp.show
dept.show
```

```
scala> emp.show
+--------+------+---------+---------+------+-------+
|staff_id|  name| position| birthday|salary|dept_id|
+--------+------+---------+---------+------+-------+
|    7369| SMITH|    CLERK|17/12/1980|   800|     20|
|    7499| ALLEN| SALESMAN|20/2/1981|  1600|     30|
|    7521|  WARD| SALESMAN|22/2/1981|  1250|     30|
|    7566| JONES|  MANAGER| 2/4/1981|  2975|     20|
|    7654|MARTIN| SALESMAN|28/9/1981|  1250|     30|
|    7698| BLAKE|  MANAGER| 1/5/1981|  2850|     30|
|    7782| CLARK|  MANAGER| 9/6/1981|  2450|     10|
|    7839|  KING|PRESIDENT|17/11/1981|  5000|     10|
|    7844|TURNER| SALESMAN| 8/9/1981|  1500|     30|
|    7900| JAMES|    CLERK| 3/12/1981|   950|     30|
|    7902|  FORD|  ANALYST| 3/12/1981|  3000|     20|
|    7934|MILLER|    CLERK|23/1/1982|  1300|     10|
|    7777|  JACK|    CLEAN| 1/11/1980|  1000|     60|
+--------+------+---------+---------+------+-------+
```

```
scala> dept.show
+-------+----------+---------+
|dept_id| dept_name|dept_city|
+-------+----------+---------+
|     10|ACCOUNTING| NEW YORK|
|     20|  RESEARCH|   DALLAS|
|     30|     SALES|  CHICAGO|
|     40|OPERATIONS|   BOSTON|
|     50|  logistics|  Seattle|
+-------+----------+---------+
```

图 5-56 使用 show() 方法查看 emp 中的数据　　图 5-57 使用 show() 方法查看 dept 中的数据

由于 emp 和 dept 本身数据量没有超过 20 条，因此使用 show() 方法会显示所有数据。

（2）show(numRows:Int)。

show(numRows:Int) 方法可以指定要显示的行数。

在 spark-shell 中输入以下命令，指定显示 emp 中前 5 行的数据，结果如图 5-58 所示。

emp.show(5)

```
scala> emp.show(5)
+--------+------+---------+----------+------+-------+
|staff_id|  name| position|  birthday|salary|dept_id|
+--------+------+---------+----------+------+-------+
|    7369| SMITH|    CLERK|17/12/1980|   800|     20|
|    7499| ALLEN| SALESMAN| 20/2/1981|  1600|     30|
|    7521|  WARD| SALESMAN| 22/2/1981|  1250|     30|
|    7566| JONES|  MANAGER|  2/4/1981|  2975|     20|
|    7654|MARTIN| SALESMAN| 28/9/1981|  1250|     30|
+--------+------+---------+----------+------+-------+
only showing top 5 rows
```

图 5-58　使用 show() 方法指定显示 emp 中前 5 行的数据

2）collect()

collect() 方法类似 Spark RDD 中的 collect() 方法。与 show() 方法不同的是，collect() 方法可以获取 DataFrame 中的所有数据，并返回一个 Array 对象。

在 spark-shell 中输入以下命令，获取 emp 中的所有数据，结果如图 5-59 所示。

emp.collect()

```
scala> emp.collect()
res4: Array[org.apache.spark.sql.Row] = Array([7369,SMITH,CLERK,17/12/1980,800
,20], [7499,ALLEN,SALESMAN,20/2/1981,1600,30], [7521,WARD,SALESMAN,22/2/1981,1
250,30], [7566,JONES,MANAGER,2/4/1981,2975,20], [7654,MARTIN,SALESMAN,28/9/198
1,1250,30], [7698,BLAKE,MANAGER,1/5/1981,2850,30], [7782,CLARK,MANAGER,9/6/198
1,2450,10], [7839,KING,PRESIDENT,17/11/1981,5000,10], [7844,TURNER,SALESMAN,8/
9/1981,1500,30], [7900,JAMES,CLERK,3/12/1981,950,30], [7902,FORD,ANALYST,3/12/
1981,3000,20], [7934,MILLER,CLERK,23/1/1982,1300,10], [7777,JACK,CLEAN,1/11/19
80,1000,60])
```

图 5-59　使用 collect() 方法获取 emp 中的所有数据

2. 过滤查询

1）where()

在 SQL 语法中，可以通过 where 条件语句对数据进行过滤。在 Spark SQL 中，同样可以使用像 SQL 一样的 where(conditionExpr:String) 条件语句对数据进行过滤，通过指定条件参数 conditionExpr 来进行查询，参数中可以使用 and 或 or。

在 spark-shell 中输入以下命令，使用 where() 过滤 emp 中薪水大于 2000 元的员工信息，结果如图 5-60 所示。

emp.where("salary > 2000").show

```
scala> emp.where("salary > 2000").show
+--------+-----+---------+----------+------+-------+
|staff_id| name| position|  birthday|salary|dept_id|
+--------+-----+---------+----------+------+-------+
|    7566|JONES|  MANAGER|  2/4/1981|  2975|     20|
|    7698|BLAKE|  MANAGER|  1/5/1981|  2850|     30|
|    7782|CLARK|  MANAGER|  9/6/1981|  2450|     10|
|    7839| KING|PRESIDENT|17/11/1981|  5000|     10|
|    7902| FORD|  ANALYST| 3/12/1981|  3000|     20|
+--------+-----+---------+----------+------+-------+
```

图 5-60　使用 where() 方法过滤信息

在上面的示例中，使用 df 的 where() 方法传入了条件表达式，对数据进行过滤，获得了薪水（salary）大于 2000 元的员工信息。

2）filter()

filter() 方法和 where() 方法类似，可以筛选出符合条件的数据。

在 spark-shell 中输入以下命令，使用 filter() 方法过滤 emp 中薪水大于 2000 元的员工信息，结果如图 5-61 所示。

emp.filter("salary > 2000").show

3. 指定列查询

1）select()

在 Spark SQL 中可以通过 select() 方法查询指定列，类似 SQL 中的 select。

在 spark-shell 中输入以下命令，查询 emp 中"name"和"salary"列的值，结果如图 5-62 所示。

emp.select("name","salary").show

图 5-61　使用 filter() 方法过滤信息

图 5-62　使用 select() 方法查询指定列

2）selectExpr()

selectExpr() 方法是 select() 方法的一个变体，可以接收一个 SQL 表达式，或对某些列做一些特殊处理。selectExpr() 方法可以通过传入 String 类型的参数来指定列调用 UDF 或者指定别名，并返回新的 DataFrame。

在 spark-shell 中输入以下命令，查询 emp 中的"name"和"salary"列，将"salary"列重命名为 xinshui，结果如图 5-63 所示。

emp.selectExpr("name","salary as xinshui").show

4. 排序操作

1）orderBy()

orderBy() 方法能够使数据按照指定的字段排序，默认按升序排序。如果需要降序排序，则可以使用 desc(" 字段名称 ") 或者 $" 字段名 ".desc 进行处理。

在 spark-shell 中输入以下命令，使用 orderBy() 方法对 emp 中的"salary"列进行降序排序，结果如图 5-64 所示。

emp.orderBy($"salary".desc).show()

```
scala> emp.selectExpr("name","salary as xinshui").show
+------+------+
| name|xinshui|
+------+------+
| SMITH|   800|
| ALLEN|  1600|
|  WARD|  1250|
| JONES|  2975|
|MARTIN|  1250|
| BLAKE|  2850|
| CLARK|  2450|
|  KING|  5000|
|TURNER|  1500|
| JAMES|   950|
|  FORD|  3000|
|MILLER|  1300|
|  JACK|  1000|
+------+------+
```

图 5-63　使用 selectExpr() 方法查询指定列

```
scala> emp.orderBy($"salary".desc).show()
+--------+------+---------+----------+------+-------+
|staff_id|  name| position|  birthday|salary|dept_id|
+--------+------+---------+----------+------+-------+
|    7839|  KING|PRESIDENT|17/11/1981|  5000|     10|
|    7902|  FORD|  ANALYST| 3/12/1981|  3000|     20|
|    7566| JONES|  MANAGER|  2/4/1981|  2975|     20|
|    7698| BLAKE|  MANAGER|  1/5/1981|  2850|     30|
|    7782| CLARK|  MANAGER|  9/6/1981|  2450|     10|
|    7499| ALLEN| SALESMAN|20/2/1981|  1600|     30|
|    7844|TURNER| SALESMAN|  8/9/1981|  1500|     30|
|    7934|MILLER|    CLERK|23/1/1982|  1300|     10|
|    7521|  WARD| SALESMAN|22/2/1981|  1250|     30|
|    7654|MARTIN| SALESMAN|28/9/1981|  1250|     30|
|    7777|  JACK|    CLEAN| 1/11/1980|  1000|     60|
|    7900| JAMES|    CLERK| 3/12/1981|   950|     30|
|    7369| SMITH|    CLERK|17/12/1980|   800|     20|
+--------+------+---------+----------+------+-------+
```

图 5-64　使用 orderBy() 方法进行排序

2）sort()

sort() 方法与 orderBy() 方法一样，也可以用于对指定列进行排序。

在 spark-shell 中输入以下命令，使用 sort() 方法对 emp 中的"salary"列进行升序排序，结果如图 5-65 所示。

emp.sort(asc("salary")).show()

```
scala> emp.sort(asc("salary")).show()
+--------+------+---------+----------+------+-------+
|staff_id|  name| position|  birthday|salary|dept_id|
+--------+------+---------+----------+------+-------+
|    7369| SMITH|    CLERK|17/12/1980|   800|     20|
|    7900| JAMES|    CLERK| 3/12/1981|   950|     30|
|    7777|  JACK|    CLEAN| 1/11/1980|  1000|     60|
|    7521|  WARD| SALESMAN|22/2/1981|  1250|     30|
|    7654|MARTIN| SALESMAN|28/9/1981|  1250|     30|
|    7934|MILLER|    CLERK|23/1/1982|  1300|     10|
|    7844|TURNER| SALESMAN|  8/9/1981|  1500|     30|
|    7499| ALLEN| SALESMAN|20/2/1981|  1600|     30|
|    7782| CLARK|  MANAGER|  9/6/1981|  2450|     10|
|    7698| BLAKE|  MANAGER|  1/5/1981|  2850|     30|
|    7566| JONES|  MANAGER|  2/4/1981|  2975|     20|
|    7902|  FORD|  ANALYST| 3/12/1981|  3000|     20|
|    7839|  KING|PRESIDENT|17/11/1981|  5000|     10|
+--------+------+---------+----------+------+-------+
```

图 5-65　使用 sort() 方法进行排序

5. 分组聚合

1）groupBy() 分组

groupBy() 方法可以根据字段进行分组操作，通过传入 String 类型的字段名或者 Column 类型的对象调用。

在 spark-shell 中输入以下命令，分别传入 String 类型的字段名和 Column 类型的对象，对 emp 中的"position（职位）"列进行分组，结果如图 5-66 所示。

emp.groupBy("position")
emp.groupBy(emp("position"))

第 5 章 电商大数据分析与探索

```
scala> emp.groupBy("position")
res13: org.apache.spark.sql.RelationalGroupedDataset = RelationalGroupedDatase
t: [grouping expressions: [position: string], value: [staff_id: int, name: str
ing ... 4 more fields], type: GroupBy]

scala> emp.groupBy(emp("position"))
res14: org.apache.spark.sql.RelationalGroupedDataset = RelationalGroupedDatase
t: [grouping expressions: [position: string], value: [staff_id: int, name: str
ing ... 4 more fields], type: GroupBy]
```

图 5-66 groupBy() 根据字段名分组

groupBy() 方法返回的是 RelationalGroupedDataset 对象，该对象提供了 max()、min()、mean()、sum()、count() 等相关的聚合函数，可以对分组后的数据进行指定的聚合操作。

在 spark-shell 中输入以下命令，先按照"position"列对 emp 对象进行分组，再计算分组中元素的平均值，可以获得不同职位的平均薪水，结果如图 5-67 所示。

emp.groupBy("position").mean("salary").show

```
scala> emp.groupBy("position").mean("salary").show
+---------+------------------+
| position|       avg(salary)|
+---------+------------------+
|    CLEAN|            1000.0|
|  ANALYST|            3000.0|
| SALESMAN|            1400.0|
|    CLERK|1016.6666666666666|
|  MANAGER|2758.3333333333335|
|PRESIDENT|            5000.0|
+---------+------------------+
```

图 5-67 使用 groupBy() 方法分组计算平均值

2）agg() 聚合

agg() 方法一般与 groupBy() 方法配合使用。agg(expers:column*) 可以传入一个或多个 Column 对象，返回 DataFrame。

在 spark-shell 中输入以下命令，对"position"列进行分组，计算各职位薪水的平均值，并命名为 average_wages，结果如图 5-68 所示。

emp.groupBy("position").agg(avg("salary").as("average_wages")).show()

```
scala> emp.groupBy("position").agg(avg("salary").as("average_wages")).show()
+---------+------------------+
| position|     average_wages|
+---------+------------------+
|    CLEAN|            1000.0|
|  ANALYST|            3000.0|
| SALESMAN|            1400.0|
|    CLERK|1016.6666666666666|
|  MANAGER|2758.3333333333335|
|PRESIDENT|            5000.0|
+---------+------------------+
```

图 5-68 agg() 聚合操作

6. 聚合函数

聚合是大数据分析领域中最常用的特性之一。Spark SQL 提供了很多常用的聚合函数，比如 sum()、count()、avg() 等。

1）count()

count() 函数可以执行以下两种操作：对指定列进行计数；使用 count(*) 或 count(1)

对所有列进行计数。

在 spark-shell 中输入以下命令，在 emp 对象中对"name"列进行计算，结果如图 5-69 所示。

emp.select(count("name")).show

2）countDistinct()

countDistinct() 函数可以计算每个组的唯一项。在 spark-shell 中输入以下命令，计算"position"列中有几个职位，结果如图 5-70 所示。

emp.select(countDistinct("position")).show

```
scala> emp.select(count("name")).show
+-----------+
|count(name)|
+-----------+
|         13|
+-----------+
```

```
scala> emp.select(countDistinct("position")).show
+-----------------------+
|count(DISTINCT position)|
+-----------------------+
|                      6|
+-----------------------+
```

图 5-69　count() 函数　　　　　　　　图 5-70　countDistinct() 函数

3）approx_count_distinct()

approx_count_distinct() 函数的功能与 countDistinct() 函数一样，但是统计的结果允许有误差。approx_count_distinct() 函数的第 2 个参数指定了可容忍的误差的最大值。

在 spark-shell 中输入以下命令，计算"position"列中有几个职位。

emp.select(approx_count_distinct("position", 0.01)).show

```
scala> emp.select(approx_count_distinct("position", 0.01)).show
+-----------------------------+
|approx_count_distinct(position)|
+-----------------------------+
|                            6|
+-----------------------------+
```

图 5-71　approx_count_distinct() 函数

由图 5-71 可知，虽然 approx_count_distinct() 函数与 countDistinct() 函数的使用效果一样，但是在处理庞大的数据集时，使用近似计算可以提高程序的执行效率。

4）sum()

sum() 函数用于计算一个数字列中的所有值的总和。

在 spark-shell 中输入以下命令，计算"salary"列的总和，结果如图 5-72 所示。

emp.select(sum("salary")).show

```
scala> emp.select(sum("salary")).show
+-----------+
|sum(salary)|
+-----------+
|      25925|
+-----------+
```

图 5-72　sum() 函数

5）sumDistinct()

sumDistinct() 函数用于计算一个数字列中不同值的总和，重复的值将不被计算。

在 spark-shell 中输入以下命令，计算 "salary" 列中不同值的总和，结果如图 5-73 所示。

emp.select(sumDistinct("salary")).show

```
scala> emp.select(sumDistinct("salary")).show
+-------------------+
|sum(DISTINCT salary)|
+-------------------+
|              24675|
+-------------------+
```

图 5-73　sumDistinct() 函数

6）avg()

avg() 函数用于计算一个数字列的平均值。

在 spark-shell 中输入以下命令，计算 "salary" 列的平均值，结果如图 5-74 所示。

emp.select(avg("salary"), (sum("salary") / count("salary"))).show

```
scala> emp.select(avg("salary"), (sum("salary") / count("salary"))).show
+-----------------+-----------------------------+
|      avg(salary)|(sum(salary) / count(salary))|
+-----------------+-----------------------------+
|1994.2307692307693|           1994.2307692307693|
+-----------------+-----------------------------+
```

图 5-74　avg() 函数

由图 5-74 可知，直接使用 avg() 函数计算的平均值与使用 sum() 函数计算的薪水总和除以 count() 函数计算的薪水数量的值相同。

7）min()、max()

在 spark-shell 中输入以下命令，统计指定列的最小值和最大值，结果如图 5-75 所示。

emp.select(min("salary"), max("salary")).show

```
scala> emp.select(min("salary"), max("salary")).show
+-----------+-----------+
|min(salary)|max(salary)|
+-----------+-----------+
|        800|       5000|
+-----------+-----------+
```

图 5-75　min() 函数、max() 函数

由图 5-75 可知，最高薪水是 5000 元，最低薪水是 800 元。

7. 连接操作

在实际工作中，有时候数据并不在同一个表中，此时需要根据业务需求，在连接两个表之后查询相关数据。DataFrame 提供了 join 功能实现连接的效果。对 DataFrame 对象 emp 和 dept 进行以下连接操作，以理解 join 连接的方法。

1）内连接

使用内连接查询操作可以通过比较运算符比较被连接列的列值来输出与连接条件匹配的数据行。

在 spark-shell 中输入以下命令，对 DataFrame 对象 emp 和 dept 进行内连接，结果如图 5-76 所示。

```
// 连接条件为 emp 中的 "dept_id" 和 dept 中的 "dept_id" 相等
emp.join(dept, emp("dept_id").equalTo(dept("dept_id")), "inner").show
// 因为 "inner" 是默认的，所以可以不指定 join 类型，这两条命令效果相同
emp.join(dept, emp("dept_id").equalTo(dept("dept_id"))).show
```

```
scala> emp.join(dept, emp("dept_id").equalTo(dept("dept_id"))).show
+--------+------+---------+---------+------+-------+-------+----------+---------+
|staff_id|  name| position| birthday|salary|dept_id|dept_id| dept_name|dept_city|
+--------+------+---------+---------+------+-------+-------+----------+---------+
|    7369| SMITH|    CLERK|17/12/1980|   800|     20|     20|  RESEARCH|   DALLAS|
|    7499| ALLEN| SALESMAN| 20/2/1981|  1600|     30|     30|     SALES|  CHICAGO|
|    7521|  WARD| SALESMAN| 22/2/1981|  1250|     30|     30|     SALES|  CHICAGO|
|    7566| JONES|  MANAGER|  2/4/1981|  2975|     20|     20|  RESEARCH|   DALLAS|
|    7654|MARTIN| SALESMAN| 28/9/1981|  1250|     30|     30|     SALES|  CHICAGO|
|    7698| BLAKE|  MANAGER|  1/5/1981|  2850|     30|     30|     SALES|  CHICAGO|
|    7782| CLARK|  MANAGER|  9/6/1981|  2450|     10|     10|ACCOUNTING| NEW YORK|
|    7839|  KING|PRESIDENT|17/11/1981|  5000|     10|     10|ACCOUNTING| NEW YORK|
|    7844|TURNER| SALESMAN|  8/9/1981|  1500|     30|     30|     SALES|  CHICAGO|
|    7900| JAMES|    CLERK| 3/12/1981|   950|     30|     30|     SALES|  CHICAGO|
|    7902|  FORD|  ANALYST| 3/12/1981|  3000|     20|     20|  RESEARCH|   DALLAS|
|    7934|MILLER|    CLERK| 23/1/1982|  1300|     10|     10|ACCOUNTING| NEW YORK|
+--------+------+---------+---------+------+-------+-------+----------+---------+
```

图 5-76 对 DataFrame 对象 emp 和 dept 进行内连接

2）左外连接

左外连接以左表为基准对数据进行连接，显示左表中的所有数据，设置右表对应的列为 null。

在 spark-shell 中输入以下命令，对 DataFrame 对象 emp 和 dept 进行左外连接。连接类型既可以是 "left_outer"，又可以是 "leftouter"。连接条件的表达式为 Seq("dept_id","dept_id")，与内连接中的 emp("dept_id").equalTo(dept("dept_id") 效果一样。结果如图 5-77 所示。

```
emp.join(dept, Seq("dept_id","dept_id"), "left_outer").show
```

```
scala> emp.join(dept, Seq("dept_id","dept_id"), "left_outer").show
+-------+-------+--------+------+---------+----------+------+----------+---------+
|dept_id|dept_id|staff_id|  name| position|  birthday|salary| dept_name|dept_city|
+-------+-------+--------+------+---------+----------+------+----------+---------+
|     20|     20|    7369| SMITH|    CLERK|17/12/1980|   800|  RESEARCH|   DALLAS|
|     30|     30|    7499| ALLEN| SALESMAN| 20/2/1981|  1600|     SALES|  CHICAGO|
|     30|     30|    7521|  WARD| SALESMAN| 22/2/1981|  1250|     SALES|  CHICAGO|
|     20|     20|    7566| JONES|  MANAGER|  2/4/1981|  2975|  RESEARCH|   DALLAS|
|     30|     30|    7654|MARTIN| SALESMAN| 28/9/1981|  1250|     SALES|  CHICAGO|
|     30|     30|    7698| BLAKE|  MANAGER|  1/5/1981|  2850|     SALES|  CHICAGO|
|     10|     10|    7782| CLARK|  MANAGER|  9/6/1981|  2450|ACCOUNTING| NEW YORK|
|     10|     10|    7839|  KING|PRESIDENT|17/11/1981|  5000|ACCOUNTING| NEW YORK|
|     30|     30|    7844|TURNER| SALESMAN|  8/9/1981|  1500|     SALES|  CHICAGO|
|     30|     30|    7900| JAMES|    CLERK| 3/12/1981|   950|     SALES|  CHICAGO|
|     20|     20|    7902|  FORD|  ANALYST| 3/12/1981|  3000|  RESEARCH|   DALLAS|
|     10|     10|    7934|MILLER|    CLERK| 23/1/1982|  1300|ACCOUNTING| NEW YORK|
|     60|     60|    7777|  JACK|    CLEAN| 1/11/1980|  1000|      null|     null|
+-------+-------+--------+------+---------+----------+------+----------+---------+
```

图 5-77 对 DataFrame 对象 emp 和 dept 进行左外连接

由图 5-77 可知，右表没有部门编号为 60 的数据，因此最后一行为空。

3）右外连接

右外连接以右表为基准对数据进行连接，显示右表中的所有数据，设置左表对应的

列为 null。

在 spark-shell 中输入以下命令，对 DataFrame 对象 emp 和 dept 进行右外连接，结果如图 5-78 所示。

```
emp.join(dept, Seq("dept_id","dept_id"), "right_outer").show
```

```
scala> emp.join(dept, Seq("dept_id","dept_id"), "right_outer").show
+-------+-------+--------+------+---------+----------+------+----------+---------+
|dept_id|dept_id|staff_id|  name| position|  birthday|salary| dept_name|dept_city|
+-------+-------+--------+------+---------+----------+------+----------+---------+
|     10|     10|    7934|MILLER|    CLERK| 23/1/1982|  1300|ACCOUNTING| NEW YORK|
|     10|     10|    7839|  KING|PRESIDENT|17/11/1981|  5000|ACCOUNTING| NEW YORK|
|     10|     10|    7782| CLARK|  MANAGER|  9/6/1981|  2450|ACCOUNTING| NEW YORK|
|     20|     20|    7902|  FORD|  ANALYST| 3/12/1981|  3000|  RESEARCH|   DALLAS|
|     20|     20|    7566| JONES|  MANAGER|  2/4/1981|  2975|  RESEARCH|   DALLAS|
|     20|     20|    7369| SMITH|    CLERK|17/12/1980|   800|  RESEARCH|   DALLAS|
|     30|     30|    7900| JAMES|    CLERK| 3/12/1981|   950|     SALES|  CHICAGO|
|     30|     30|    7844|TURNER| SALESMAN|  8/9/1981|  1500|     SALES|  CHICAGO|
|     30|     30|    7698| BLAKE|  MANAGER|  1/5/1981|  2850|     SALES|  CHICAGO|
|     30|     30|    7654|MARTIN| SALESMAN| 28/9/1981|  1250|     SALES|  CHICAGO|
|     30|     30|    7521|  WARD| SALESMAN| 22/2/1981|  1250|     SALES|  CHICAGO|
|     30|     30|    7499| ALLEN| SALESMAN| 20/2/1981|  1600|     SALES|  CHICAGO|
|     40|     40|    null|  null|     null|      null|  null|OPERATIONS|   BOSTON|
|     50|     50|    null|  null|     null|      null|  null|  logistics|  Seattle|
+-------+-------+--------+------+---------+----------+------+----------+---------+
```

图 5-78 对 DataFrame 对象 emp 和 dept 进行右外连接

由图 5-78 可知，左表没有 "OPERATIONS" 和 "LOGISTICS" 两个部门的数据，因此最后两行为空。

4）全外连接

全外连接在等值连接的基础上添加左表和右表中的未匹配数据。

在 spark-shell 中输入以下命令，对 DataFrame 对象 emp 和 dept 进行全外连接，结果如图 5-79 所示。

```
emp.join(dept, Seq("dept_id","dept_id"), "outer").show
```

```
scala> emp.join(dept, Seq("dept_id","dept_id"), "outer").show
+-------+-------+--------+------+---------+----------+------+----------+---------+
|dept_id|dept_id|staff_id|  name| position|  birthday|salary| dept_name|dept_city|
+-------+-------+--------+------+---------+----------+------+----------+---------+
|     20|     20|    7369| SMITH|    CLERK|17/12/1980|   800|  RESEARCH|   DALLAS|
|     20|     20|    7566| JONES|  MANAGER|  2/4/1981|  2975|  RESEARCH|   DALLAS|
|     20|     20|    7902|  FORD|  ANALYST| 3/12/1981|  3000|  RESEARCH|   DALLAS|
|     40|     40|    null|  null|     null|      null|  null|OPERATIONS|   BOSTON|
|     10|     10|    7782| CLARK|  MANAGER|  9/6/1981|  2450|ACCOUNTING| NEW YORK|
|     10|     10|    7839|  KING|PRESIDENT|17/11/1981|  5000|ACCOUNTING| NEW YORK|
|     10|     10|    7934|MILLER|    CLERK| 23/1/1982|  1300|ACCOUNTING| NEW YORK|
|     50|     50|    null|  null|     null|      null|  null|  logistics|  Seattle|
|     60|     60|    7777|  JACK|    CLEAN| 1/11/1980|  1000|      null|     null|
|     30|     30|    7499| ALLEN| SALESMAN| 20/2/1981|  1600|     SALES|  CHICAGO|
|     30|     30|    7521|  WARD| SALESMAN| 22/2/1981|  1250|     SALES|  CHICAGO|
|     30|     30|    7654|MARTIN| SALESMAN| 28/9/1981|  1250|     SALES|  CHICAGO|
|     30|     30|    7698| BLAKE|  MANAGER|  1/5/1981|  2850|     SALES|  CHICAGO|
|     30|     30|    7844|TURNER| SALESMAN|  8/9/1981|  1500|     SALES|  CHICAGO|
|     30|     30|    7900| JAMES|    CLERK| 3/12/1981|   950|     SALES|  CHICAGO|
+-------+-------+--------+------+---------+----------+------+----------+---------+
```

图 5-79 对 DataFrame 对象 emp 和 dept 进行全外连接

由图 5-79 可知，全外连接会保留左、右表中的所有数据，并且用 null 填充缺失的数据。

5）左半连接

左半连接（Left Semi-join）的行为类似内连接类型，连接后的数据集只包含匹配的行。

在 spark-shell 中输入以下命令，对 DataFrame 对象 emp 和 dept 进行半连接，结果如图 5-80 所示。

emp.join(dept, Seq("dept_id","dept_id"), "left_semi").show

图 5-80　对 DataFrame 对象 emp 和 dept 进行半连接

6）交叉连接

如果在执行 join 连接时不指定参数，那么得到的是笛卡儿积结果。

在 spark-shell 中输入以下命令，对 DataFrame 对象 emp 和 dept 进行交叉连接（又称 Cartesian Join，笛卡儿连接），结果如图 5-81 所示。

// 使用交叉连接并显示结果
emp.crossJoin(dept).show
// 以下命令效果相同
emp.join(dept).show

图 5-81　对 DataFrame 对象 emp 和 dept 进行交叉连接

5.2.2　数据分析探索

基于前述的知识与方法，本节将对"在线销售订单"数据集文件"orders.csv"和

"order-details.csv"进行分析和探索，解决情景导入中提出的问题。在进行处理分析之前，依旧是先将两份数据集文件上传至 HDFS 中，再加载数据集到 RDD 中，最后加载到 DataFrame 中进行分析。

1. 数据准备

案例分析——
在线销售订单
数据分析

（1）上传"orders.csv"和"order-details.csv"文件到 HDFS 中。

在 master 节点中输入以下命令，数据集上传结果如图 5-82 所示：

[root@master ~]# hdfs dfs -put /root/data/Chapter5/orders.csv /Chapter5/
[root@master ~]# hdfs dfs -put /root/data/Chapter5/order-details.csv /Chapter5/

```
[root@master ~]# hdfs dfs -put /root/data/Chapter5/orders.csv /Chapter5/
2022-10-27 12:49:38,671 INFO sasl.SaslDataTransferClient: SASL encryption trust
 check: localHostTrusted = false, remoteHostTrusted = false
[root@master ~]# hdfs dfs -put /root/data/Chapter5/order-details.csv /Chapter5/
2022-10-27 12:49:54,908 INFO sasl.SaslDataTransferClient: SASL encryption trust
 check: localHostTrusted = false, remoteHostTrusted = false
[root@master ~]#
```

图 5-82　上传数据集到 HDFS 中

（2）加载数据集到 RDD 和 DataFrame 中。

在 spark-shell 的 paste 模式下输入以下命令，加载 DataFrame，结果如图 5-83 所示。

```
// 定义"orders.csv"文件的路径
val filePath1 = "/Chapter5/orders.csv"
// 加载订单数据到 DataFrame 中
val orders = spark.read.option("header","true").
        option("inferSchema","true").
        csv("/Chapter5/orders.csv")
// 查看 orders 中前 5 行的数据
orders.show(5)
// 查看 orders 的 schema 模式
orders.printSchema()

// 定义"order-details.csv"文件的路径
val filePath2 = "/Chapter5/order-details.csv"
// 加载订单详细数据到 DataFrame 中
val orders_details = spark.read.option("header","true").
        option("inferSchema","true").
        csv("/Chapter5/order-details.csv")
// 查看 orders 中前 5 行的数据
orders_details.show(5)
// 查看 orders 的 schema 模式
orders_details.printSchema()
```

```
scala> :paste
// Entering paste mode (ctrl-D to finish)

//定义 "orders.csv" 文件的路径
val filePath1 = "/Chapter5/orders.csv"
// 加载订单数据到DataFrame中
val orders = spark.read.option("header","true").
            option("inferSchema","true").
            csv("/Chapter5/orders.csv")
//查看orders中前5行的数据
orders.show(5)
//查看orders的schema模式
orders.printSchema()

// Exiting paste mode, now interpreting.

+---------+----------+-------+----------+------+
| order_id|order_date|cust_id| full_name|  city|
+---------+----------+-------+----------+------+
|100354678| 2020/10/1|  60124|Titus.Jani|Vinson|
|100354678| 2020/10/1|  60124|Titus.Jani|Vinson|
|100354680| 2020/10/1|  60124|Titus.Jani|Vinson|
|100354680| 2020/10/1|  60124|Titus.Jani|Vinson|
|100367357|2020/11/13|  60124|Titus.Jani|Vinson|
+---------+----------+-------+----------+------+
only showing top 5 rows

root
 |-- order_id: string (nullable = true)
 |-- order_date: string (nullable = true)
 |-- cust_id: integer (nullable = true)
 |-- full_name: string (nullable = true)
 |-- city: string (nullable = true)

filePath1: String = /Chapter5/orders.csv
orders: org.apache.spark.sql.DataFrame = [order_id: string, order_date: string ... 3 more fields]

scala> :paste
// Entering paste mode (ctrl-D to finish)

//定义 "order-details.csv" 文件的路径
val filePath2 = "/Chapter5/order-details.csv"
// 加载订单详细数据到DataFrame中
val orders_details = spark.read.option("header","true").
            option("inferSchema","true").
            csv("/Chapter5/order-details.csv")
//查看orders中前5行的数据
orders_details.show(5)
//查看orders的schema模式
orders_details.printSchema()

// Exiting paste mode, now interpreting.

+---------+-----------------+-----------+------+-------------+---------------+
| order_id|              sku|qty_ordered| price|     category|discount_percent|
+---------+-----------------+-----------+------+-------------+---------------+
|100354678| oasis_Oasis-064-36|         21|  89.9|Men's Fashion|            0.0|
|100354678|      Fantastic_FT-48|         11|  19.0|Men's Fashion|            0.0|
|100354680|      mdeal_DMC-610-8|          9| 149.9|Men's Fashion|            0.0|
|100354680| oasis_Oasis-061-36|          9|  79.9|Men's Fashion|            0.0|
|100367357|MEFNAR59C38B6CA08CD|          2|  99.9|Men's Fashion|            0.0|
+---------+-----------------+-----------+------+-------------+---------------+
only showing top 5 rows

root
 |-- order_id: string (nullable = true)
 |-- sku: string (nullable = true)
 |-- qty_ordered: integer (nullable = true)
 |-- price: double (nullable = true)
 |-- category: string (nullable = true)
 |-- discount_percent: double (nullable = true)

filePath2: String = /Chapter5/order-details.csv
orders_details: org.apache.spark.sql.DataFrame = [order_id: string, sku: string ... 4 more fields]
```

图 5-83　加载数据集到 RDD 和 DataFrame 中

2. 问题分析

1）统计此数据集中每个客户的下单数量

每个客户的下单情况数据在 orders 中。orders 中的一条数据就是一个订单。因此，只需要对其中的"full_name（客户姓名）"列进行 groupBy() 分类并使用 count() 函数计算总数即可。还可以使用 sort() 方法，对客户下单数量进行降序排序，获得下单最多的客户的信息。

在 spark-shell 中输入以下命令，计算每个客户的下单数量，结果如图 5-84 所示。

val custordernum = orders.groupBy("full_name").count().sort($"count".desc).show(10)

```
scala> val custordernum = orders.groupBy("ful
l_name").count().sort($"count".desc).show(10)

+---------------+-----+
|      full_name|count|
+---------------+-----+
|   Gonzalez.Joel| 2524|
|    Bailes.Eulah|  707|
|        Melo.Liz|  608|
|    Braddy.Percy|  436|
| Beebe.Hortencia|  397|
|       Hohn.Saul|  329|
|       Nally.Ned|  306|
| Glines.Cathrine|  304|
|Matthies.Kenton|  285|
|   Jesse.Alfonso|  277|
+---------------+-----+
only showing top 10 rows

custordernum: Unit = ()
```

图 5-84　每个客户的下单数量

由图 5-84 可知，在所有客户中，Gonzalez.Joel 下单的数量最多，是此商店的一位忠实客户。

2）每个城市的订单

此问题的分析思路与上一题相同，只需将分析的列替换为"city（城市）"列即可。在 spark-shell 中输入以下命令，计算每个城市的订单量，结果如图 5-85 所示。

orders.groupBy("city").count().sort($"count".desc).show(10)

```
scala> orders.groupBy("city").count().sort($"count".desc).show(10)
+-------------+-----+
|         city|count|
+-------------+-----+
|       Dekalb| 2525|
|   Washington| 2008|
|New York City| 1391|
|      Houston| 1250|
|      El Paso| 1019|
|      Atlanta|  999|
|       Dallas|  838|
|  Springfield|  817|
|       Albany|  726|
|    Kittanning|  716|
+-------------+-----+
only showing top 10 rows
```

图 5-85　每个城市的订单量

由图 5-85 可知，Dekalb 的下单量最多，其次是 Washington。各市的订单量数据可以为商家的备货方案提供建议。

3）每个客户的消费总额

针对此问题，需要对数据做一些转换。通过观察"order-details.csv"中的数据可以发现，每一条记录中都包含商品的单价和销售数量，并且有些商品是有折扣的。有的商品销售记录来自同一个订单，即同一个客户购买的。所以要计算每个客户的年消费总额，首先要计算每条记录的消费数量，接着根据订单编号进行求和，获得每个客户的消费总额。

（1）计算每条记录中商品的实际销售金额。

实际销售金额 = 商品价格 * 已订购数量 - 商品价格 * 已订购数量 * 折扣百分比

在 spark-shell 中输入以下命令,结果如图 5-86 所示。

```
val orders_details1 = orders_details.select($"order_id",(($"price" * $"qty_ordered") - ($"price" * $"qty_ordered") * $"discount_percent"*0.01).as("expend_amount"))
// 查看计算结果
orders_details1.show(5)
```

```
scala> val orders_details1 = orders_details.select($"order_id",(($"price" * $"qty_ordered
") - ($"price" * $"qty_ordered") * $"discount_percent"*0.01).as("expend_amount"))
orders_details1: org.apache.spark.sql.DataFrame = [order_id: string, expend_amount: doubl
e]

scala> orders_details1.show()
+---------+------------------+
| order_id|     expend_amount|
+---------+------------------+
|100354678|            1887.9|
|100354678|             209.0|
|100354680|1349.1000000000001|
|100354680|             719.1|
|100367357|             199.8|
|100367357|              79.8|
|100367360|              95.2|
|100354677|              98.0|
|100354677|             270.0|
|100354677|            1099.8|
|100356116|    519.90000000237|
|100358724|   529.8999999979139|
|100403034|     429.98743999848|
|100403034|    532.4216600295999|
|100403034|    117.79089999988|
|100403077|     429.98743999848|
|100403077|    532.4216600295999|
|100403077|    117.79089999988|
|100403411|     429.98743999848|
|100403411|    532.4216600295999|
+---------+------------------+
only showing top 20 rows
```

图 5-86 每条记录中商品的实际销售金额

(2)按照订单编号统计每个订单的总金额。

首先,根据订单编号"order_id"使用 groupBy() 方法对获得的数据 orders_details1 进行分类;然后,使用 sum() 函数对"expend_amount"列进行求和并重命名为 OrderTotal,从而获得每个订单的总金额。

在 spark-shell 中输入以下命令,结果如图 5-87 所示。

```
// 计算每个订单的总金额
val ordertotal = orders_details1.groupBy("order_id").agg(sum("expend_amount").as("OrderTotal"))
// 查看输出结果,对"OrderTotal"列数据保留 2 位小数,并对"order_id"列升序排序
ordertotal.select($"order_id",bround($"OrderTotal",2)).sort("order_id").show(5)
```

```
scala> val ordertotal = orders_details1.groupBy("order_id").agg(sum("expend_amount").as("
OrderTotal"))
ordertotal: org.apache.spark.sql.DataFrame = [order_id: string, OrderTotal: double]

scala> ordertotal.select($"order_id",bround($"OrderTotal",2)).sort("order_id").show(5)
+---------+-------------------+
| order_id|bround(OrderTotal, 2)|
+---------+-------------------+
|100354677|             1467.8|
|100354678|             2096.9|
|100354679|              613.0|
|100354680|             2068.2|
|100354681|             2199.8|
+---------+-------------------+
only showing top 5 rows
```

图 5-87 每个订单的总金额

第 5 章 电商大数据分析与探索

（3）进行等值内连接，并增加订单总金额。

客户编号、客户名字等信息均在 orders 中，每张订单的总金额数据在 ordertotal 中，要想获得每个客户的消费总额，就需要连接相关的数据表。但是在 orders 中有许多重复的 order_id，因此需要对其进行去重处理。

在 spark-shell 中输入以下命令，去重结果如图 5-88 所示。

```
// 使用 dropDuplicates() 对 orders 中的重复行进行删除
val ordersdrop = orders.dropDuplicates()
```

```
scala> val ordersdrop = orders.dropDuplicates()
ordersdrop: org.apache.spark.sql.Dataset[org.apache.spark.sql.Row] = [order_id: string, order_date: string ... 3 more fields]
```

图 5-88　对 orders 中的重复行进行删除

对 ordersdrop 和 ordertotal 中的 order_id 进行等值内连接，并选择 ordersdrop 中的"order_id""order_date""cust_id""full_name""order_date""city"列，对 ordertotal 中的"OrderTotal"列进行输出，保留两位小数并重命名为 Total_consumption。在 spark-shell 中输入以下命令，结果如图 5-89 所示。

```
val orders1 = ordersdrop.join(ordertotal, ordersdrop("order_id").equalTo(ordertotal("order_id")), "inner").select(ordersdrop("order_id"),ordersdrop("order_date"),ordersdrop("cust_id"),ordersdrop("full_name"),ordersdrop("city"),bround(ordertotal("OrderTotal"),2).alias("Total_consumption"))
```

```
// 查看处理结果，对"Total_consumption"列降序排序输出
orders1.sort($"Total_consumption".desc).show()
```

```
scala> val orders1 = ordersdrop.join(ordertotal, ordersdrop("order_id").equalTo(ordertot
al("order_id")), "inner").select(ordersdrop("order_id"),ordersdrop("order_date"),ordersd
rop("cust_id"),ordersdrop("full_name"),ordersdrop("city"),bround(ordertotal("OrderTotal"
),2).alias("Total_consumption"))
orders1: org.apache.spark.sql.DataFrame = [order_id: string, order_date: string ... 4 mo
re fields]

scala> orders1.sort($"Total_consumption".desc).show()
+---------+----------+-------+---------------+----------------+-----------------+
| order_id|order_date|cust_id|      full_name|            city|Total_consumption|
+---------+----------+-------+---------------+----------------+-----------------+
|100458711| 2021/3/24|  89637|    Hosea.Brock|     Head Waters|         304848.6|
|100510800| 2021/5/11|  17987|   Davison.Jeff|       Spearfish|        202525.18|
|100517032|  2021/6/1| 104056|   Duby.Eustolia|       Indianola|         151372.6|
|100518707|  2021/6/6| 106389|   Fung.Miquel |        Oak Park|        148996.22|
|100518623|  2021/6/6| 106389|   Fung.Miquel |        Oak Park|        122521.68|
|100518737|  2021/6/6| 106389|   Fung.Miquel |        Oak Park|        121998.42|
|100517409|  2021/6/3| 105009|     Lear.Elmo |      East Point|         105716.0|
|100542627| 2021/7/19| 110399|      Knop.Sue |     Great Falls|         101118.2|
|100533924| 2021/6/25| 107853|    Hord.Adolfo|            Ross|         100658.4|
|100510802| 2021/5/11| 104056|   Duby.Eustolia|       Indianola|          97800.0|
|100562387| 2021/9/30| 113474|   Appell.Hilton|         Detroit|          97781.7|
|100543286| 2021/7/23| 111127|   Champion.Tena|Colorado Springs|          89397.3|
|100542580| 2021/7/19| 111057| Jauregui.Bianca|   New Hyde Park|          88648.4|
|100542605| 2021/7/19| 111057| Jauregui.Bianca|   New Hyde Park|          88648.4|
|100533792| 2021/6/25| 107853|    Hord.Adolfo|            Ross|          88073.4|
|100542597| 2021/7/19| 111057| Jauregui.Bianca|   New Hyde Park|          86378.4|
|100419053|2020/12/27|  78546|Toothaker.Renaldo|       Hightown|          84804.3|
|100542579| 2021/7/19| 111057| Jauregui.Bianca|   New Hyde Park|          83748.4|
|100544042| 2021/7/25| 111438|   Hickson.Kris|          Canton|          81457.8|
|100535400| 2021/6/29| 109032|     Boris.Chia|         Howland|          80132.0|
+---------+----------+-------+---------------+----------------+-----------------+
only showing top 20 rows
```

图 5-89　ordersdrop 和 ordertotal 等值内连接结果

由图 5-89 可知每个订单对应的客户信息、订单消费总额及订单的详细信息。由于同一个客户拥有若干个不同的订单，因此需要对同一个客户的订单消费总额进行求和，

获得每个客户的消费总额。

在 spark-shell 中输入以下命令,获得每个客户的消费总额,结果如图 5-90 所示。

```
// 计算每个客户的消费总额
val orders2 = orders1.groupBy("cust_id","full_name").agg(sum("Total_consumption").as("Total_consumption1"))
// 对消费总额降序排序
orders2.sort($"Total_consumption1".desc).show
```

图 5-90 每个客户的消费总额

4)客户的平均订单消费额

由图 5-89 可知,在 orders1 中,同一个用户可能有多个不同的订单。先对客户编号"cust_id"列和客户名字"full_name"列进行 groupBy() 分类,再对"Total_consumption"列进行 avg() 平均,即可得到客户的平均订单消费额。

在 spark-shell 中输入以下命令,获取客户的平均订单消费额,结果如图 5-91 所示。

```
val ordersavg = orders1.groupBy("cust_id","full_name").agg(avg("Total_consumption").as("orderAvg"))
// 降序查看平均订单消费额
ordersavg.sort(col("orderAvg").desc).show()
```

图 5-91 客户的平均订单消费额

通过以上分析和探索，可以获得客户在此商店的下单数量、消费总额、平均订单消费额情况等信息。此外，还可以根据商品编号、年份、月份等信息进行数据转换、计算分析，以获取更多有价值的信息。

脑图小结

本章介绍了使用 Spark SQL 进行数据分析与探索的方法。通过"女装电子商务评论"数据分析和"在线销售订单"数据分析两个任务讲解数据准备、清洗、转换、分析、输出、查询过程中用到的 DataFrame 操作方法。通过以下脑图小结，助力学习者掌握和巩固相关知识。

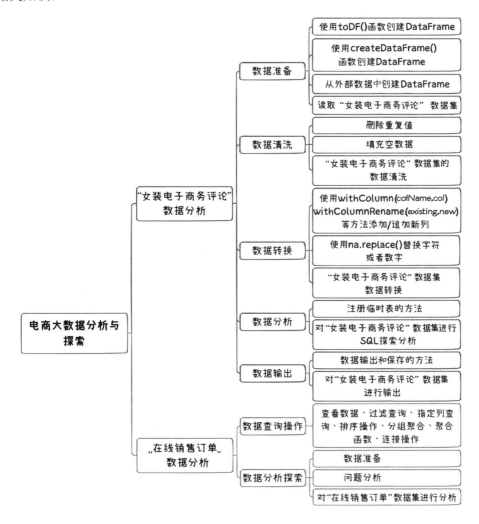

章节练习

1. 现有一份某化妆品线上商店的客户行为数据集文件"User-behavior.csv"（数据集文件在本章的数据文件夹中），数据集中的字段及含义如下表所示。请根据以下需求，使用 Spark SQL 相关知识对数据集进行操作。

字　　段	含　　义
event_time	客户行为时间
event_type	客户行为类型
product_id	化妆产品 ID
brand	化妆产品品牌
price	产品价格
user_id	客户 ID

（1）请读取该文件并创建 DataFrame，查看前 5 行数据。
（2）在该数据集中，有多少个客户对该线上商店进行了浏览？
（3）在该数据集的时间段内，所有商品属于多少个品牌分类？
（4）在这段时间中，该线上商店每天卖出了多少商品？
（5）请根据客户行为数据，统计这段时间客户"view（浏览）"、"cart（加入购物车）"、"remove_from_cart（从购物车删除）"和"purchase（购买）"这几种行为的占比。
（6）除了无品牌，哪种品牌的化妆品被购买的数量最多？

2. 利用题 1 的数据，使用 Spark SQL 中 DataFrame 的各种操作方法对数据进行查询和分析，完成以下任务。
（1）了解各个产品的销售额情况，并查看销售额最高的产品。
（2）获取被浏览次数最多的产品 ID。
（3）查看每个品牌的销售额情况，以及销售额最高的化妆品品牌。

第 6 章

Zeppelin 数据可视化

任务 6.1　Zeppelin 安装与部署

Zepplin
安装部署

情境导入

数据可视化可以使数据信息更清晰、形象地展示给数据分析人员,也为数据分析结果的汇报和呈现提供了有效途径。大数据分析及可视化的方法和工具有很多,Zeppelin 就是其中一种,它是数据分析人员使用较多的一个工具。

Zeppelin 是 Apache 基金会下的一个开源框架。它是一个基于 Web 的 Notebook,提供交互式数据分析和可视化;还是一个高性能、高可用的分布式键 – 值对存储平台,以高性能、大集群为目标。Zeppelin 后台支持接入多种数据处理引擎,如 Spark、Hive 等;支持多种语言,如 Scala(Apache Spark)、Python(Apache Spark)、Spark SQL、Hive、Markdown、Shell 等。

Zeppelin 可视化工具具有数据提取、数据发掘、数据分析、数据可视化展示与合作等功能。它并不局限于 Spark SQL 的查询,任何后端语言的输出都可以被识别并可视化呈现。我们可以在 Notebook 中动态地创建一些输入格式,以实现不同的可视化效果。此外,还可以通过共享 Notebook 的 URL 实现协同开发,Zeepelin 可视化工具会实时同步 Notebook 的任何改动。本节将讲解 Zeppelin 的安装与部署。

学习目标和要求

知识与技能目标

1. 能下载、安装、配置 Zeppelin 可视化工具。
2. 能在 Zeppelin 可视化工具中配置 Spark 解释器。

素质目标

1. 增强团队的协同意识。
2. 具有通过信息技术进行主动学习的能力，逐渐养成不断学习的习惯并提高适应发展的能力。

6.1.1 下载安装包

Zeppelin 可视化工具的安装包可以直接通过其官方网站下载。在官网中选择"zeppelin-0.10.1-bin-all.tgz"安装包进行下载，如图 6-1 所示。

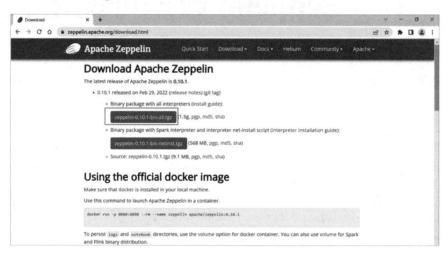

图 6-1　下载 Zeppelin 安装包

在下载完成之后，将安装包上传到 Hadoop 集群中进行安装配置。将 Zeppelin 安装包上传到 master 节点的"root"目录中，如图 6-2 所示。

图 6-2　上传安装包到 master 节点中

6.1.2 安装配置

1. 解压缩并重命名目录

（1）在 master 节点中输入以下命令，解压缩结果如图 6-3 所示。

将安装包解压缩到 "/usr/local/" 目录中
[root@master ~]# tar -zxf zeppelin-0.10.1-bin-all.tgz -C /usr/local/

图 6-3　解压缩 Zeppelin 安装包

（2）重命名文件目录，结果如图 6-4 所示。

[root@master ~]# cd /usr/local/
[root@master src]# mv zeppelin-0.10.1-bin-all/ zeppelin

图 6-4　重命名 Zeppelin 文件目录

2. 修改配置文件

（1）在 master 节点中输入以下命令，查找并复制"zeppelin-env.sh.template"模板文件，将其重命名为 zeppelin-env.sh，结果如图 6-5 所示。

[root@master local]# cd zeppelin/
[root@master zeppelin]# ls
[root@master zeppelin]# cd conf/
[root@master conf]# ls
[root@master conf]# cp zeppelin-env.sh.template zeppelin-env.sh
[root@master conf]# ls

图 6-5　复制 Zeppelin 配置模板文件

（2）修改"zeppelin-env.sh"配置文件。

在 master 节点中输入以下命令：

[root@master conf]# vi zeppelin-env.sh

在文件最后添加以下配置内容：

export JAVA_HOME=/usr/local/jdk1.8.0_281
export SPARK_HOME=/usr/local/spark
export HADOOP_CONF_DIR=/usr/local/hadoop/etc/hadoop
export ZEPPELIN_ADDR=192.168.128.130
export ZEPPELIN_PORT=9090

在配置内容中添加了 Java、Spark 的安装目录，Hadoop 配置文件目录，Zeppelin 的地址和端口，如图 6-6 所示。

```
export JAVA_HOME=/usr/local/jdk1.8.0_281
export SPARK_HOME=/usr/local/spark
export HADOOP_CONF_DIR=/usr/local/hadoop/etc/hadoop
export ZEPPELIN_ADDR=192.168.128.130
export ZEPPELIN_PORT=9090
```

图 6-6　修改 Zeppelin 配置文件

3. Zeppelin 控制台界面功能

在 master 节点中输入以下命令，启动 Zeppelin，操作结果如图 6-7 所示。

[root@master conf]# cd /usr/local/zeppelin/
[root@master zeppelin]# bin/zeppelin-daemon.sh start

```
[root@master ~]# cd /usr/local/zeppelin/
[root@master zeppelin]# bin/zeppelin-daemon.sh start
Zeppelin start                                              [  OK  ]
[root@master zeppelin]#
```

图 6-7　启动 Zeppelin

在启动 Zeppelin 之后，通过浏览器打开 Zeppelin 控制台，如图 6-8 所示。

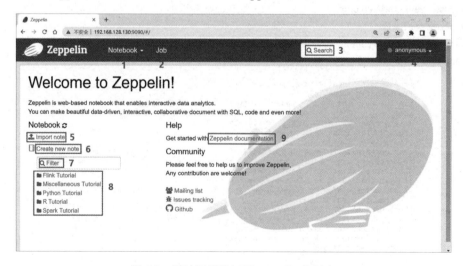

图 6-8　通过浏览器打开 Zeppelin 控制台

第6章 Zeppelin 数据可视化

在 Zeppelin 控制台中可以进行许多操作，图 6-8 中各序号标注部分的功能说明如下。

（1）全局 Notebook 管理菜单：包含 6、7、8 的功能，在 Zeppelin 的任意界面中均可快速打开此菜单，实现查看、切换、创建 Notebook 的功能。

（2）运行任务管理：查看所有运行的任务，包括运行完成、正在运行、失败、已取消等状态。

（3）全局搜索：不仅能搜索 Notebook，还能搜索 Notebook 中的代码。

（4）用户相关的管理设置：包括解释器管理、Notebook 仓库管理、授权管理、查看系统配置项、用户注销等设置。

（5）导入 Notebook：从文件中导入一个 Notebook 应用。

（6）创建 Notebook：创建新的 Notebook 应用，在创建时可指定应用名称和默认代码解释器。

（7）按名称筛选 Notebook。

（8）Notebook 列表：显示所有 Notebook 应用，支持多级目录组织。

（9）Zeppelin 帮助文档链接。

4. Zeppelin 中的 Spark 解释器配置

Zeppelin 支持 Scala、Python、Flink、Spark SQL、Hive、JDBC、Markdown、Shell 等多种解释器。本书将使用 Spark 解释器进行配置。

（1）点击界面右上角的"anonymous"按钮，在下拉列表中选择"Interpreter"选项，打开解释器配置界面，如图 6-9 所示。

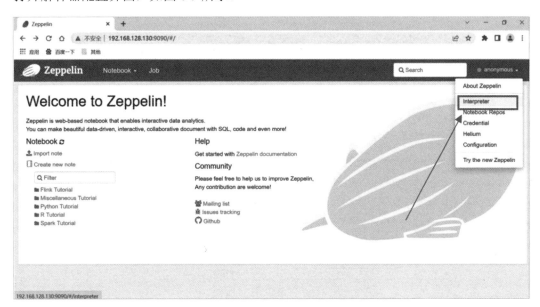

图 6-9 打开解释器配置界面

（2）在搜索框里输入"spark"搜索解释器，点击右边的"edit"按钮，进入 Spark 解释器配置界面，如图 6-10 所示。

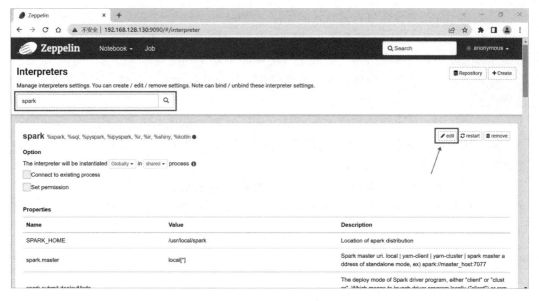

图 6-10　进入 Spark 解释器配置界面

修改 Spark 解释器配置。在配置选项中，SPARK_HOME 需要根据实际的 Spark 安装路径进行配置。spark.master 需要根据 Spark 部署类型进行填写，可以是 local 模式、yarn-client 模式、standalone 模式等，本任务采用的是 standalone 模式，如图 6-11 所示。在完成修改之后，将界面拖曳到最后并点击"save"按钮进行保存。

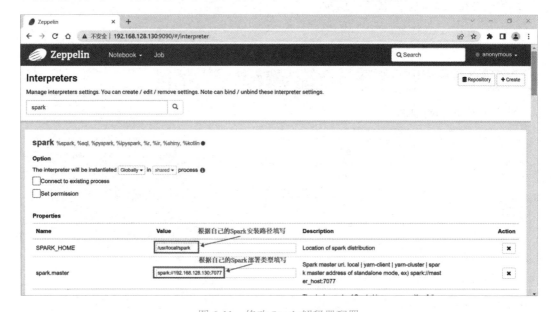

图 6-11　修改 Spark 解释器配置

6.1.3 测试运行 Zeppelin

1. 测试自带的 Spark Basic Features 案例

点击菜单栏中的"Notebook"按钮,在下拉列表中选择"Spark Basic Features"选项,本任务对一份银行数据进行分析及可视化,如图 6-12 所示。

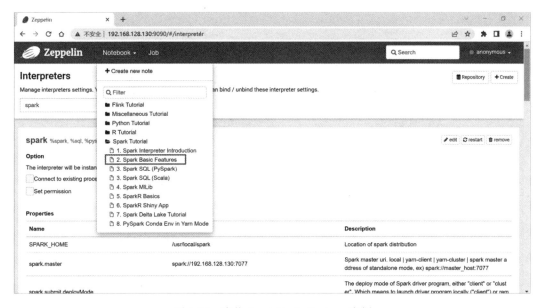

图 6-12　查找 Spark Basic Features 案例

点击"运行"按钮,运行 Spark Basic Features 案例,如图 6-13 所示。

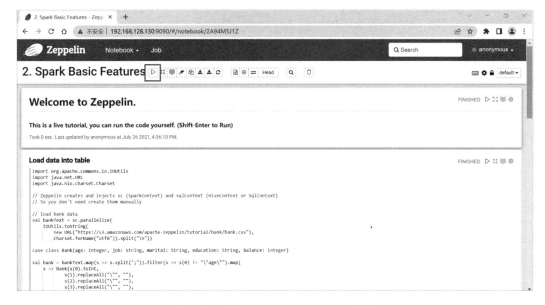

图 6-13　运行 Spark Basic Features 案例

运行后的数据图表能正确显示，图 6-14 所示为安装成功。

图 6-14　查看 Spark Basic Features 案例运行结果

2. 测试运行自编代码

以"2.2.3 Spark 集群测试"中计算矩形面积为例进行测试运行。

（1）回到 Zeppelin 控制台中，点击"Create new note"按钮，创建一个新的 Notebook，如图 6-15 所示。

图 6-15　创建 Notebook

（2）在弹出的界面中填写新文件的路径和名称，选择解释器，此处选择默认的 Spark 解释器即可，如图 6-16 所示。点击"Create"按钮，创建文件。

第 6 章　Zeppelin 数据可视化

图 6-16　填写文件名称及选择解释器

（3）在新建的 Notebook 界面中运行以下命令：

val r = 5
val pi = 3.14
val area = pi * r * r
println(" 面积为：" + area)

将上述命令复制到命令输入栏中，点击"运行"按钮，运行后的计算结果会显示在界面下方，如图 6-17 所示。在此界面中执行命令的好处是，命令输入栏中的命令在执行之后依旧可以进行修改，并再次运行。

图 6-17　测试运行代码

3. Notebook 界面功能

Notebook 界面是数据分析人员执行主要操作的界面，所有的分析功能都能在这里实现。该界面的主要功能如图 6-18 所示。

图 6-18 Notebook 界面的主要功能

各序号标注部分的功能说明如下。

（1）运行应用中所有的 Paragraph。

（2）是否显示代码编辑区域。

（3）是否显示输出结果区域。

（4）清空输出结果。

（5）复制 Notebook。

（6）以 zpln 格式导出 Notebook，此文件格式是 Zeppelin 的默认格式。

（7）以 ipynb 格式导出 Notebook，".ipynb" 文件是在使用 JupyterNotebook 编写 Python 程序时产生的文件。

（8）重新加载 Notebook。

（9）版本控制，可以选择添加提交消息以保存当前状态。

（10）对比修订。

（11）搜索代码。

（12）删除 Notebook。

（13）快捷键列表。

（14）绑定解释器，可以绑定使用的解释器。

（15）设置权限。

（16）布局选项，可以选择界面的不同布局方式。

（17）显示 Paragraph 的运行状态。

（18）运行 Paragraph。

（19）是否显示本 Paragraph 的代码编辑器。

（20）是否显示本 Paragraph 的输出结果。

（21）设置更丰富的功能，支持控制 Paragraph 的宽度、字体大小、上下移动、显示标题（show title）、显示行号、删除等。

（22）代码编辑区域。

（23）输出结果显示区域。

Notebook 提供了快捷键功能，用户可以方便地进行增加、删除、移动、运行、复制和粘贴操作，常见快捷键功能如表 6-1 所示。

表 6-1　Notebook 的快捷键功能

快　捷　键	功　　能
Shift + Enter	运行 Paragraph
Ctrl + Shift+ UP	运行当前以上所有 Paragraph（不包括）
Ctrl + Shift + DOWN	运行当前以下所有 Paragraph（包括）
Ctrl + Alt + C	取消
Ctrl + P	向上移动光标
Ctrl + N	向下移动光标
Ctrl + Alt + D	删除 Paragraph
Ctrl + Alt + A	在上面插入新 Paragraph
Ctrl + Alt + B	在下面插入新 Paragraph
Ctrl + Shift + C	插入 Paragraph 的副本
Ctrl + Alt + K	上移 Paragraph
Ctrl + Alt + J	下移 Paragraph
Ctrl + Alt + R	启用 / 禁用运行 Paragraph
Ctrl + Alt + O	切换到输出
Ctrl + Alt + E	切换到编辑
Ctrl + Alt + M	切换行号
Ctrl + Alt + T	切换标题
Ctrl + Alt + L	清除输出
Ctrl + Alt + W	链接此 Paragraph
Ctrl + Shift + -	减小 Paragraph 宽度
Ctrl + Shift + +	增大 Paragraph 宽度
Ctrl + .	自动完成
Ctrl + K	切断线路
Ctrl + S	在代码中搜索
Ctrl + A	将光标移到开头
Ctrl + E	将光标移到末尾
Ctrl + Alt + F	在代码中查找

任务 6.2　"女装电子商务评论"数据可视化

情境导入

在第 5 章中，我们利用 Spark SQL 在 spark-shell 中输入命令，对"女装电子商务评论"数据集进行了分析。但是通过此方法分析获得的结果都是二维数据表，数据特征不够直观、生动，不利于总结、分析和汇报。因此，此女装线上商店负责人希望能对数据进行可视化，以更好地进行决策。

通过"任务 6.1 Zeppelin 安装与部署"的操作，使用 Zeppelin 可视化工具就可以满足此要求。在 Zeppelin 可视化工具的 Notebook 中使用解释器，可以将查询的结果以柱状图、散点图、面积折线图等常见的方式进行可视化展示。本任务将对"女装电子商务评论"数据集进行可视化分析展示。

学习目标和要求

知识与技能目标
1. 掌握在 Zeppelin 可视化工具中使用 Spark SQL 注册视图的方法。
2. 能通过 Zeppelin 可视化工具执行 Spark SQL 命令实现数据可视化。

素质目标
1. 培养能从不同角度分析问题的发散思维。
2. 激发创新意识和学习热情。

6.2.1 加载数据注册视图

案例分析——"女装电子商务评论"数据可视化

在进行数据准备和分析之前，启动 Hadoop 集群、Spark 集群和 Zeppelin。在 Zeppelin 控制台中点击 "Create new note" 按钮，创建一个新的 Notebook，并命名为 clothing_reviews，如图 6-19 所示。

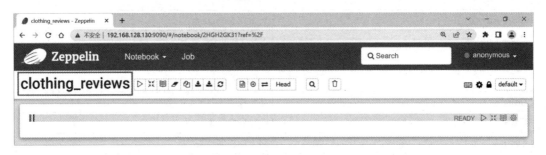

图 6-19　创建一个新的 Notebook

1. 数据准备

读取 HDFS 中的 CSV 文件，通过自定义 schema 将其转换为 DataFrame。

```
import org.apache.spark.sql._
import org.apache.spark.sql.types._

// 读取 CSV 文件存储目录
var file = "/Chapter5/Clothing-Reviews.csv"

// 自定义一个 schema，定义字段名称及数据类型
var fields= List(
    StructField("order_id", IntegerType, true),
```

StructField("clothing_id", StringType, true),
  StructField("age", IntegerType, true),
  StructField("review_text",StringType, true),
  StructField("rating",IntegerType, true),
  StructField("recommended_IND",StringType, true),
  StructField("positive_feedback_count",IntegerType, true),
  StructField("class_name",StringType, true),
)
val schema = StructType(fields)

// 读取 CSV 文件，在创建 DataFrame 时指定 schema
var reviews = spark.read.option("header","true").schema(schema).csv(file)

将上述命令复制到 clothing_reviews 的命令输入栏中并运行，创建 DataFrame，输出结果如图 6-20 所示。

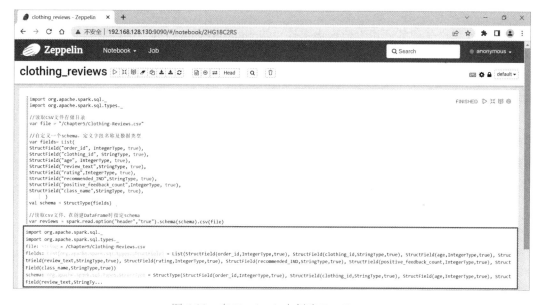

图 6-20　在 Notebook 中创建 DataFrame

输入以下命令，打印显示 DataFrame 的 schema 信息，输出结果如图 6-21 所示。

reviews.printSchema()

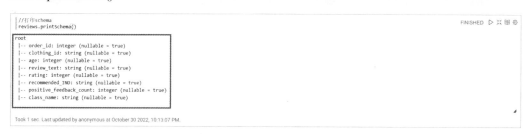

图 6-21　在 Notebook 中查看 schema 信息

输入以下命令，显示 DataFrame 的前 20 条数据信息，输出结果如图 6-22 所示。

reviews.show()

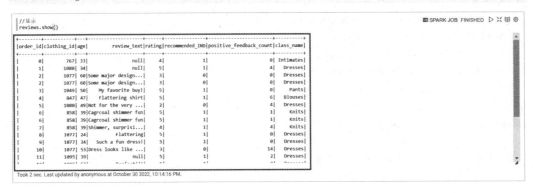

图 6-22　在 Notebook 中显示 DataFrame 的前 20 条数据

2. 数据清洗

对"女装电子商务评论"数据集文件"Clothing-Reviews.csv"进行数据清洗，删除重复的数据行和有空数据的行。

输入以下命令，数据清洗前后的数据量如图 6-23 所示，与"5.1.2 数据清洗"中执行命令的结果一致。

```
// 删除重复的数据行
val reviews1=reviews.dropDuplicates()
// 统计查看对 reviews1 处理后的数据数量
reviews1.count()
// 删除有空数据的行
val reviews2=reviews1.na.drop()
// 统计对 reviews2 进行处理后的数据数量
reviews2.count()
```

图 6-23　在 Notebook 中查看数据清洗前后的数据量

输入以下命令，查看并删除重复行和有空数据的行，数据清洗后的数据如图 6-24 所示。

reviews2.show()

第 6 章 Zeppelin 数据可视化

图 6-24　在 Notebook 中查看数据清洗后的数据

3. 数据转换

对"女装电子商务评论"数据集文件"Clothing-Reviews.csv"进行数据转换,将"recommended_IND(是否推荐)"列中的 1 替换为"推荐",0 替换为"不推荐"。此转换是对上述数据清洗后的 reviews2 数据进行的操作。

输入以下命令,输出结果如图 6-25 所示。

val reviews3=reviews2.na.replace("recommended_IND",Map("0" -> " 不推荐 ","1" -> " 推荐 "))
reviews3.show

图 6-25　在 Notebook 中查看数据转换结果

4. 创建视图

使用 createOrReplaceTempView() 方法创建本地的临时视图 clothing_reviews。
输入以下命令,输出结果如图 6-26 所示。

rviews3.createOrReplaceTempView("clothing_reviews")

图 6-26　在 Notebook 中创建视图

6.2.2 执行 SQL 数据可视化

在使用 Spark SQL 命令时,必须在第一行输入"%sql",目的是告诉 Zeppelin 的解释器,后续输入的命令是 Spark SQL 命令。

(1)对创建好的"clothing_reviews"视图进行操作和分析。首先了解该线上商店的客户年龄分布。查看 40 岁以下不同年龄段的客户人数。

输入以下命令,输出结果如图 6-27 所示。

```
%sql
select age,count(age) as total_nu
    from clothing_reviews
    where age<40
    group by age
    order by age
```

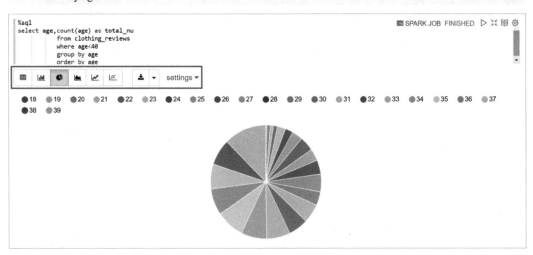

图 6-27 不同年龄段的客户人数可视化结果

由图 6-27 可知,在 Zeppelin 中,数据可视化可以通过表格、柱状图、饼图、面积折线图、折线趋势图、散点图来呈现。还可以将分析获得的数据以 CSV 或者 TSV 的格式下载到电脑上。此外,在"settings"下拉列表中可以对数据的 keys、values 等值进行设置,以观察不同状态下的数据情况。

(2)查看销量最高的服装分类。

输入以下命令,输出结果如图 6-28 所示。

```
%sql
select clothing_id,count(1) num
    from clothing_reviews
    group by clothing_id
    order by num desc
```

第 6 章 Zeppelin 数据可视化

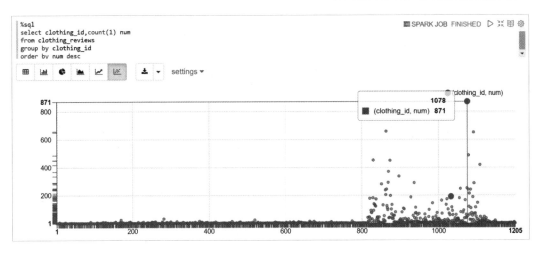

图 6-28 销量最高的服装分类可视化结果

（3）统计不同评分的"推荐"和"不推荐"，以及 recommended_IND 的数量情况。输入以下命令，输出结果如图 6-29 所示。

%sql
select recommended_IND,rating, count(1)
　　from clothing_reviews
　　group by rating,recommended_IND

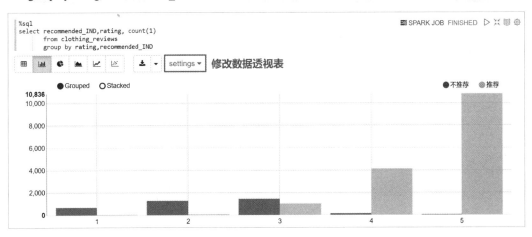

图 6-29 不同评分的"推荐"和"不推荐"可视化结果

由图 6-29 可知，在 Zeppelin 中，可以将各评分等级及"推荐"和"不推荐"的数量呈现在一张表格中。五星好评的商品推荐数量是最多的。四星好评的商品推荐数量锐减，三星、二星、一星的推荐数量更少，不推荐的数量有所增多。此结果分布符合常规认知。

图 6-29 所示的表格还可以在"settings"下拉列表中修改为数据透视表，以不同的形式呈现。在 Zeppelin 中，可以采用简单的拖曳方式对数据进行聚合并生成数据透视表，还可以创建包括求和、计数、平均、最小值、最大值等多个值的集合的数据透视表。

209

在 Zeppelin 数据透视表中，keys 表示横坐标轴要显示的数据，当多个 key 并存时，所有 key 会做笛卡儿积运算；values 表示纵坐标要显示的数据，它与 keys 对应，当存在多个 value 时，一个 key 对应多个 value（key 与 value 的关系是 1：n）。value 的可选属性有 max、min、sum、count、avg。groups 可以根据 group 的内容对 value 再次分组。

在图 6-30 中，将 rating 设置为 keys，recommended_IND 设置为 groups 分组，count(1) 求和数量设置为 values，呈现的透视数据表与图 6-29 相同。选择"values"列表框中的"count(1) SUM"选项，在下拉列表中选择需要的计算方式进行计算展示。

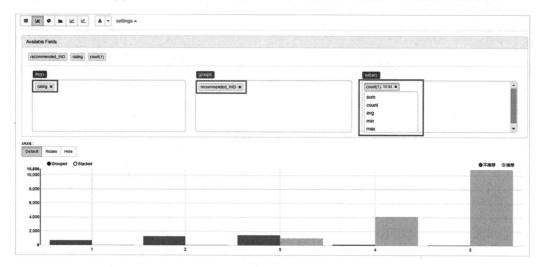

图 6-30 设置数据透视表（1）

在图 6-31 中，将"recommended_IND"拖曳到"keys"列表框中，将"rating"拖曳到"groups"列表框中，透视数据表会以另一种聚合形态展示。由此可见在 Zeppelin 中可以自定义显示图表信息，数据可视化交互体验更强。

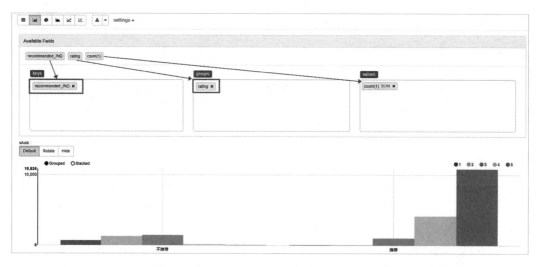

图 6-31 设置数据透视表（2）

（4）查询"不推荐"的订单分别属于哪些服装分类，以及各种服装分类的不推荐数量。

输入以下命令,输出结果如图 6-32 所示。

```sql
%sql
select class_name, count(1) as classnum
from clothing_reviews
where recommended_IND = '不推荐'
group by class_name
```

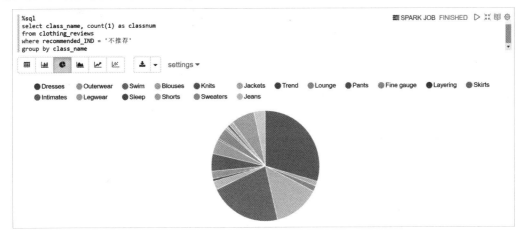

图 6-32　不推荐的服装分类情况可视化结果

(5)现在,商店负责人想知道本商店五星好评最多的服装分类是什么,评分最低的服装分类是什么。对于此问题,只需要统计和分析不同评分等级的服装编号 clothing_id 即可。此处的服装编号代表具体某件服装。

我们可以创建一个下拉列表,使数据可视化更具交互性。

输入以下命令,输出结果如图 6-33 所示。在"stars"下拉列表中可以选择 1～5 的评分等级,结果也会随之更新,从而动态查看各个评分等级的服装分类数量情况。

```sql
%sql
select clothing_id,count(1) as stars_num
from clothing_reviews
where rating=${stars=rating,5|4|3|2|1}
group by clothing_id
```

由图 6-33 可知,当鼠标移动到散点图中位置最高的点时,会弹出此点的具体信息,可以获知五星好评最多的服装分类编号为 1078。

我们也可以创建一个接收输入值的文本框,在文本框中输入不同的值,进行数据的分析观察。

输入以下命令,输出结果如图 6-34 所示。可以看到,在文本框中,可以输入 1～5 的评分等级,对于输入的不同等级,输出的散点图也会随之改变。

```sql
%sql
select clothing_id,count(1) as stars_num
from clothing_reviews
where rating=${stars=5}
```

group by clothing_id

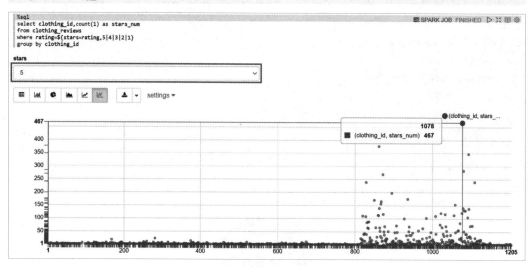

图 6-33　可视化下拉列表设置

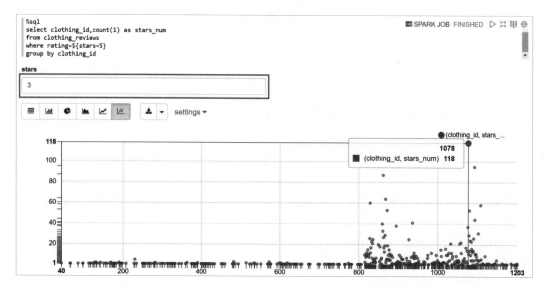

图 6-34　可视化文本框设置

任务 6.3　"在线销售订单"数据可视化

情境导入

通过对"任务 6.2 '女装电子商务评论'数据可视化"的学习，我们已经掌握了在 Zeppelin 可视化工具中使用 Spark SQL 创建视图并执行 Spark SQL 命令实现数据可视化的方法。现在商店负责人要求对"在线销售订单"数据进行可视化分析。

学习目标和要求

知识与技能目标
灵活编写 Spark SQL 命令，应用 Zeppelin 实现数据可视化分析。

素质目标
具有举一反三的知识迁移能力。

执行 Spark SQL 数据可视化

1. 在 Zeppelin 的 Notebook 中加载 DataFrame

在 Notebook 的命令输入栏输入以下命令，加载 DataFrame。

```
// 定义"orders.csv"文件的路径
val filePath1 = "/Chapter5/orders.csv"
// 加载订单数据到 DataFrame 中
val orders = spark.read.option("header","true").
        option("inferSchema","true").
        csv("/Chapter5/orders.csv")
// 查看 orders 中前 5 行的数据
orders.show(5)
// 查看 orders 的 schema 模式
orders.printSchema()


// 定义"order-details.csv"文件的路径
val filePath2 = "/Chapter5/order-details.csv"
// 加载订单详细数据到 DataFrame 中
val orders_details = spark.read.option("header","true").
        option("inferSchema","true").
        csv("/Chapter5/order-details.csv")
// 查看 orders 中前 5 行的数据
orders_details.show(5)
// 查看 orders 的 schema 模式
orders_details.printSchema()
```

2. 使用 createOrReplaceTempView() 方法创建本地的临时视图 orders_tmp

```
orders.createOrReplaceTempView("orders_tmp")
```

3. 统计此数据集中每个客户的下单数量，了解客户忠诚度

```
%sql
select full_name,count(full_name) as total_orders
    from orders_tmp
    group by full_name
```

order by total_orders desc
limit 10

在执行以上命令之后，输出结果如图 6-35 所示，客户的下单数量一目了然。

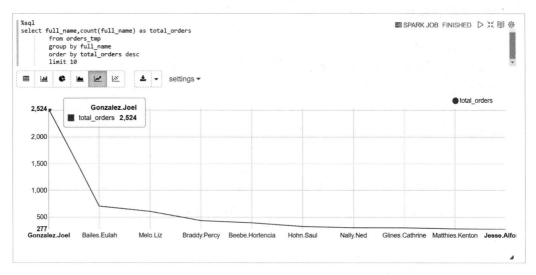

图 6-35　每个客户的下单数量可视化结果

4. 查看每个城市的订单，为商家备货方案提供建议

%sql
select city,count(city) as city_orders
　　from orders_tmp
　　group by city
　　order by city_orders desc
　　limit 10

在执行以上命令之后，输出结果如图 6-36 所示，各个城市的下单数量一目了然。

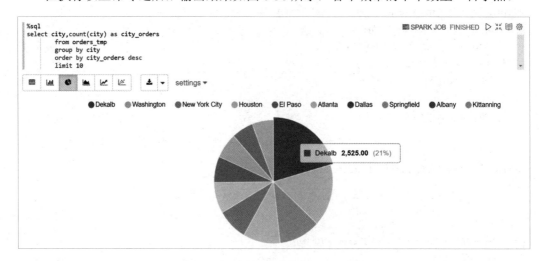

图 6-36　各个城市的下单数量可视化结果

5. 分析并计算每个客户的消费总额

为了针对不同的客户推荐合适的商品，需要获得客户的消费能力情况。因此需要分析并计算每个客户的消费总额。此处分析思路不再赘述，仅罗列操作步骤。

以下命令均在 Notebook 的命令输入栏中输入。

（1）数据处理。

计算 orders_details 中每行商品的实际销售金额。

val orders_details1 = orders_details.select($"order_id",(($"price" * $"qty_ordered") - ($"price" * $"qty_ordered") * $"discount_percent"*0.01).as("expend_amount"))

计算每张订单的消费总额。

val ordertotal = orders_details1.groupBy("order_id").agg(sum("expend_amount").as("OrderTotal"))

使用 dropDuplicates() 方法，删除 orders 中的重复行。

val ordersdrop = orders.dropDuplicates()

对 ordersdrop 和 ordertotal 进行等值内连接。

val orders1 = ordersdrop.join(ordertotal, ordersdrop("order_id").equalTo(ordertotal("order_id")), "inner").select(ordersdrop("order_id"),ordersdrop("order_date"),ordersdrop("cust_id"),ordersdrop("full_name"),ordersdrop("city"),bround(ordertotal("OrderTotal"),2).alias("Total_consumption"))

计算每个客户的消费总额。

val orders2 = orders1.groupBy("cust_id","full_name").agg(sum("Total_consumption").as("Total_consumption1"))

（2）创建视图。

创建视图 orders2_tmp。

orders2.createOrReplaceTempView("orders2_tmp")

（3）可视化展示。

对客户姓名和消费总额进行可视化展示，结果如图 6-37 所示。

%sql
select cust_id,full_name,Total_consumption1
from orders2_tmp
order by Total_consumption1 desc
limit 10

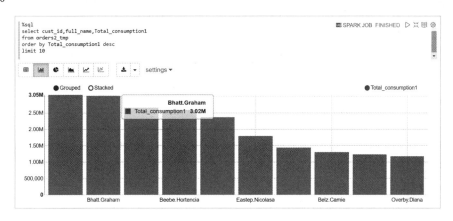

图 6-37 每个客户的消费总额可视化结果

脑图小结

本章介绍了基于 Web 的 Notebook 数据可视化工具 Zeppelin。通过详细的操作步骤，介绍了安装与部署 Zeppelin 的方法，以及 Zeppelin 控制台的具体功能。利用"女装电子商务评论"数据可视化和"在线销售订单"数据可视化两个任务，介绍了使用 Zeppelin 创建图表、操作数据透视表的方法。通过以下脑图小结，助力学习者掌握巩固相关知识。

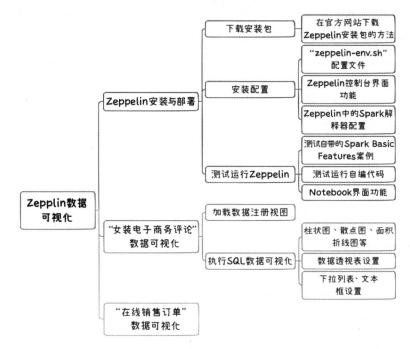

章节练习

对第 5 章的"章节练习"案例进行可视化，具体要求如下。

（1）根据客户行为数据，统计表中时间段内客户"view（浏览）"、"cart（加入购物车）"、"remove_from_cart（从购物车删除）"和"purchase（购买）"这几种行为的占比，并可视化展示。

（2）统计各个品牌化妆品的浏览、加购物车、从购物车删除、购买的数据情况，通过数据透视表展示。

（3）统计各个品牌化妆品的产品数量。

（4）了解各个产品的销售额情况。

（5）查看各个品牌的销售额。

（6）查看各个产品被浏览的次数。